The Strength of Materials & Structures

John Anderson

Copyright © BiblioBazaar, LLC

BiblioBazaar Reproduction Series: Our goal at BiblioBazaar is to help readers, educators and researchers by bringing back in print hard-to-find original publications at a reasonable price and, at the same time, preserve the legacy of literary history. The following book represents an authentic reproduction of the text as printed by the original publisher and may contain prior copyright references. While we have attempted to accurately maintain the integrity of the original work(s), from time to time there are problems with the original book scan that may result in minor errors in the reproduction, including imperfections such as missing and blurred pages, poor pictures, markings and other reproduction issues beyond our control. Because this work is culturally important, we have made it available as a part of our commitment to protecting, preserving and promoting the world's literature.

All of our books are in the "public domain" and many are derived from Open Source projects dedicated to digitizing historic literature. We believe that when we undertake the difficult task of re-creating them as attractive, readable and affordable books, we further the mutual goal of sharing these works with a larger audience. A portion of Bibliobazaar profits go back to Open Source projects in the form of a donation to the groups that do this important work around the world. If you would like to make a donation to these worthy Open Source projects, or would just like to get more information about these important initiatives, please visit www.bibliobazaar.com/opensource.

THE

STRENGTH OF MATERIALS

AND STRUCTURES.

PART I.

THE STRENGTH OF MATERIALS, AS DEPENDING ON THEIR QUALITY, AND AS ASCERTAINED BY TESTING-APPARATUS.

PART II.

THE STRENGTH OF STRUCTURES, AS DEPENDING ON THE FORM AND ARRANGEMENT OF THEIR PARTS, AND ON THE MATERIALS OF WHICH THEY ARE CONSTRUCTED.

BY

SIR JOHN ANDERSON,

C.E. LL.D. F.R.S.E.

Late Superintendent of Machinery to the War Department; formerly Lecturer on Applied Mechanics at the Royal Military Academy, Woolwich; at the Royal Engineer Establishment, Chatham; and at the Royal School of Naval Architecture, South Kensington.

ELEVENTH EDITION.

LONDON:

LONGMANS, GREEN, AND CO.

AND NEW YORK : 15 EAST 16th STREET.

1892.

All rights reserved.

PREFACE.

THIS ELEMENTARY TREATISE is divided into two distinct parts.

The First Part treats of the natural properties of various materials employed in construction, more especially in regard to their strength and elasticity, and their adaptation for particular practical purposes in the arts. The object of this portion of the work is to describe the qualities and characteristics of materials, so far as they are of importance to the engineer, or are exhibited in the results of experiments made with the testing-machine.

An acquaintance with the natural properties of materials, as forming part of the great field of applied mechanics, is indispensably necessary for the young mechanic or engineer, who desires to be something more than an artisan. That real knowledge, which consists in understanding the materials which he handles, and in a familiarity with their points of agreement or difference, both in regard to elasticity and strength, cannot fail to give a charm to his daily duty.

As it cannot be expected that all will have the opportunity of making experiments for themselves, the first three chapters are devoted to the testing of materials, and to the practical manipulation of a testing-machine. In these chapters reference is made to the physical properties of some common materials, by which the student will be able to comprehend the nature of such experimental investigations, and the labour and care needed in order to arrive at true results ; and it is hoped that he will find the subject treated in such a clear and simple manner, that he may understand it without much difficulty.

The fourth, fifth, sixth, seventh, and eighth chapters refer more especially to cast iron, wrought iron, steel, copper, alloys, and timber, and are intended to describe their qualities and leading peculiarities.

The ninth, tenth, and eleventh chapters treat, more generally, of the resistance of materials to torsion, shearing, and punching, and to transverse strains, conjoined with impact and vibration.

The first eleven chapters, therefore, have regard to the nature of materials, and the remaining six chapters—which constitute the second part of the volume—are devoted to the strength of structures, when made of the materials previously treated of.

In the Second Part, the student will learn the correct forms which must be given to the various structures in order to obtain the requisite strength, and likewise the best arrangement of materials, as depending on their respective properties, so that by the practical application of correct principles, the maximum of strength may be attained with the minimum of weight and cost.

The experiments most frequently referred to, and which are quoted as 'Woolwich experiments,' have all been made in the Royal Arsenal for various purposes during the past eighteen years, and chiefly with the American testing-machine, which is described in the third chapter, the only exception being certain experiments, to ascertain the strength of ropes under various conditions, which were carried out with a hydraulic testing-machine, recently transferred from Her Majesty's Dockyard to the Royal Arsenal.

The author has taken pains to ensure accuracy; still, when so many figures have been necessarily transcribed several times, more particularly in the Tables, some errors may probably exist. In this part of the work Mr. Charles Topple and Mr. George Cuthbert have both rendered valuable assistance; the former, more especially, in regard to the chapters on structures, where calculations were concerned. Most of the examples given are taken from actual works, which have passed through the author's hands during the past few years.

The sources from whence the results of other experiments have been drawn, or from which extracts have been made, are generally quoted. Reference has very frequently been made to the Blue Book, containing the Report of the Commissioners appointed to enquire into the Application of Iron to Railway Structures.

WOOLWICH: *August* 1872.

CONTENTS.

PART I.

ON THE STRENGTH OF MATERIALS EMPLOYED IN CONSTRUCTION.

CHAPTER I.

ON SOME OF THE PHYSICAL PROPERTIES OF MATERIALS.

Fitness of Materials for Special Purposes. Rigidity only apparent. Elasticity. Hooke's Law. Solids, Plastic Substances, Liquids and Gases. Permanent Set of Solids. Cases in which the Effects of Elasticity need to be counteracted . . . . . PAGE 1—9

CHAPTER II.

ON THE EXPERIMENTAL TESTING OF MATERIALS.

Need of Accuracy. Several Specimens should be tested. Stretching of Long and Short Specimens. Force to be applied in Axis of Specimen. Effect of Time. Limit of Elasticity. Form of Specimens. Testing Machines . . . . 9—15

CHAPTER III.

ON A MACHINE FOR TESTING THE STRENGTH AND ELASTICITY OF MATERIALS.

Woolwich Machine. Balance of Machine. Application of Load. Construction of Specimen Holders. Behaviour of Specimens. Flow of Solids. Compressive Resistance. Transverse Resistance. Torsion. Deformation . . . . 15—29

x   *Contents.*

CHAPTER IV.
CAST IRON.

Production of Cast Iron. Influence of Carbon on Cast Iron. Mixing Cast Iron. Limit of Elasticity. Tenacity. Selection and Testing by Founder. Testing Girders. Woolwich Experiments. Experiments on 10-feet Bars. Compressive Strength. Experiments on 50-feet Bars. Woolwich Compressive Experiments. Remelted Iron. Strength at various Temperatures . . PAGE 30—45

CHAPTER V.
WROUGHT IRON.

Production of Wrought Iron. Characteristics. Conditions determining Quality. Tenacity of Bars and Plates. Strength at high and low Temperatures. Appearance of Fracture. Effect of Vibration. Annealing. Safe Load. Red short and Cold short Iron. Limit of Elasticity. Behaviour under Test of hard and ductile Specimens. Strength of Welds. Experiments on 50-feet Bars. Modulus of Elasticity. Compressive Resistance . . . . 46—67

CHAPTER VI.
STEEL.

Production of Steel. Bessemer Steel. Effect of working. Whitworth's Treatment. Mild Steel. Tempering in Oil. Strength of Steel. Tempering Barrels for Guns. Experiments on Steel. Compressive Resistance. Siemens Steel. Elimination of Phosphorus in Crude Cast Iron by Gilchrist-Thomas Process. Processes of Steel Manufacture. Experiments on specimens of Siemens Steel. Composition. Relation of Tensile Strength to amount of Carbon. Extension and Elasticity. Experiments with soft Bessemer Steel. Torsion and Twisting. Tensile Strain . . . . 68—80

CHAPTER VII.
ON COPPER AND OTHER METALS, AND THEIR ALLOYS.

COPPER. Its Tenacity. Forging. Effect of Exposure. Specific Gravity. Melting Point. Wiredrawing. Casting. Use of Phosphorus. Experiments on Tenacity. BRONZE AND GUN-METAL. Specific Gravity. Fusibility. Manufacture. Experiments on Gun-metal. Resistance to Tension and Compression. BRASS. Manufacture. Specific Gravity. Melting Point. Influence of Lead. MUNTZ METAL. Tensile Strength. STERRO-METAL. ALUMINIUM BRONZE. BELL-METAL. BABBITT'S METAL. Melting Point of Alloys. Additional Experiments. Manganese-bronze. Preparation. Useful Properties. Tests of Manganese-bronze by Tensile Strain and Torsion . . . . . PAGE 81—106

CHAPTER VIII.

TIMBER.

Variability in Quality. Density. Conditions influencing Quality. Seasoning. Laws of Shrinkage. Ash. Beech. Elm. Pine and Fir. Hornbeam. Mahogany. Oak. Teak. Tensile and Compressive Resistance of Timber. Resistance to Shearing. Transverse Strength . . . . 107—127

CHAPTER IX.

TRANSVERSE STRENGTH OF IRON AND RESISTANCE TO IMPACT.

Breaking Weight and Deflection of Bars. Effect of Impact. Transverse Flexure. Effect of repeated Deflections . . 128—136

CHAPTER X.

RESISTANCE TO TORSION AND SHEARING.

Shafting. Law of Resistance to Twisting. Wrought and Cast Iron Shafting. Cast Steel and Copper under Torsion. Torsional Stiffness. Sudden Variations of Section. Experiments with Bars. Rupture by Torsion. Torsive Resistance of different Materials. Strength and Stiffness of Shafting. Resistance to Shearing and Punching. Gradual Shearing by inclined Instruments. Shearing Action on Links. Punching and Drilling . . . . . 136—155

CHAPTER XI.

ON THE IMPORTANCE OF UNIFORMITY OF SECTIONAL AREA.

Surplus Material in parts a source of weakness. Screw Bolts. Chains. Proof Strains for Chains. Rules for Strength of Chains. Strength of Ropes. Rope Slings . . . . . . 155—162

PART II.

ON THE STRENGTH OF STRUCTURES.

CHAPTER XII.

BEAMS AND GIRDERS.

Mathematics useful. Dependence on Formulæ. Beams and Girders. Relative Strength as depending on Mode of Application of Load and Support. Strength as affected by Arrangement of Material. Neutral Surface. Tension and Compression in Beams. Flanged Girders. Rectangular Beams. Tubular Beams. Beams of uniform strength. Constants for Strength and Deflection of Beams. Cast and Wrought Iron Girders. Deflection of Beams. Resilience of Beams . . . . PAGE 163—193

CHAPTER XIII.

ON THE STRENGTH OF GEARING.

Teeth of Wheels. Strength of Teeth. Flanging Wheel Teeth. Wear of Teeth. Wrought Iron Wheels. Malleable Cast Iron Wheels. Cast Steel Wheels. Power transmitted by Wheels. Strength of Screws. Cutting Instruments . . . 193—201

CHAPTER XIV.

ON THE STRENGTH OF LONG COLUMNS.

Modes in which Columns yield. Short Columns. Euler's Theory. Hodgkinson's Experiments. Practical Deductions. Wrought Iron Columns. Cast Iron Columns. Strength of Rectangular and Cylindrical Tubes. Steel Shear Poles . . . . 201—212

CHAPTER XV.

ON THE STRENGTH OF CRANES AND ROOF TRUSSES AS EXAMPLES OF COMPLEX STRUCTURES.

Mode of dealing with Complex Structures. Bracing. Triangular arrangement of parts. Common Roof Truss. Trussed or Braced Girder. Stresses on Hydraulic Crane. Strength of Sheer Legs. Construction and Calculation of Stress on Members of a Steam Crane. Crane Shafts. Crane Chain. Crane Foundations. Concrete Foundations. Cylinder Foundation. Pier-head Foundation . . . . PAGE 212—246

CHAPTER XVI.

STRENGTH OF RIVETED STRUCTURES—STEAM BOILERS, ETC.

Strength of Boiler Plates. Of Joints in Plates. Rivets. Proportions of Joints. Oval Rivets. Double-cover Plates. Single and Double Riveting. Plates with thick Edges. Diagonal Joints. Strength of Boilers. Resistance to Collapse. Provision for Expansion and Contraction. Flues. Hydraulic Test. Strength of Cylindrical Boiler without Flues. Strength of Boiler with Flues. Tubular Boiler. Firebox Stays. Butt and Lap Joints. Welded Joints. Resistance of Tubes to collapse. Fairbairn's Experiments. Elliptical Tubes. Effect of Length on Resistance to internal Pressure. Small Tube Boilers . . . . 247—276

CHAPTER XVII.

ON STRUCTURES SUBJECT TO INTERNAL PRESSURE.

Cast Iron Pipes. Pipes for High Pressures. Iron Tanks. Strengthening by Wrought Iron Hoops. Thick Cylinders. Barlow's Theory. Gun Structures. American System of Casting Guns. Strength as depending on Exterior Form. Hydraulic Press Cylinders. Chilled Shot. Built-up Guns. Construction of Hoops. Armstrong's System. Longitudinal Strain on Guns. Need of Mass in the Breech. Strengthening of Cast Iron Guns. Introduction of a Lining. General Conclusions . . . . 277—302

PART I.

ON THE STRENGTH OF MATERIALS EMPLOYED IN CONSTRUCTION.

CHAPTER I.

ON SOME OF THE PHYSICAL PROPERTIES OF MATERIALS.

IN ORDER to understand that branch of applied mechanics which treats of the strength of materials, it is first of all necessary that the student should possess a precise knowledge of those physical properties, on which the constructive value of a material and its adaptation for given circumstances depend. The fitness and reliability of materials for special purposes is by no means, exclusively, a question of strength, but is contingent on hardness, stiffness, toughness, malleability, and other inherent properties, which result from the conditions of freedom or restraint existing amongst their constituent molecules. In judging of the suitability of a material for special duties, it is further necessary to know its powers of endurance under the action either of forces tending to abrade it, or of frequently repeated loading, or of vibration and impact, or of varied changes of temperature, as the case may be.

Our knowledge of the different kinds of materials now chiefly employed in connection with practical operations, has been extended by the publication of a large mass of valuable data, obtained in the numerous experimental investigations of scientific men, during the last hundred years, and particularly during the last quarter of a century; more especially is this the case with the metals, cast iron, wrought iron, steel, copper, and brass. Hence, notwithstanding many discrepancies which occasionally present themselves, there are a great number of well ascertained facts and many definite laws for the guidance of those engaged in construction.

To the young beginner, almost every kind of material with which he comes into contact appears to present a different appearance and character from that which it really possesses, or which it will appear to possess when he knows more about it. Thus, it may happen that a substance which, at first, seemed a hard, solid, rigid, continuous mass, is, in reality, soft and porous, and even movable in its internal structure; that every description of metal or wood is yielding and elastic; that there is no such condition as absolute permanence of form, but that every material body suffers displacement of its parts under the application of external forces, being *distorted* or *deformed*, being extended, or compressed, or deflected, or twisted. The constituent molecules of some solids may even be made to slide over or amongst each other, may be spread out into a flat web, may be gathered into a more compact form, or may be crushed, to an extent dependent only on the straining force which is applied.

The apparent fixity and rigidity of solids is unreal, and this fact lies at the foundation of the whole subject; hence one of the most remarkable features for the student to observe, at the outset, is the unfailing mobility of the internal structure of materials on the application of sufficient stress. The naked eye may not be able to detect the movement, because our perceptions are not sufficiently acute to notice the minute change that takes place, but there is a change following every application of force. Hence, it is found necessary to resort to testing instruments of various kinds, in order to apply great straining forces and to make the precise movements of the mass visible. For example, by the use of a testing machine, it is found that a bar of wrought iron, one square inch in section, will be elongated $\frac{1}{10000}$th part of its length by the weight of a single ton, and will continue to stretch with every additional ton, until rupture takes place; or, to give another example of mobility, the change of temperature between the extremes of summer and winter is sufficient to expand, or contract, a wrought iron bar to the extent of $\frac{1}{3000}$th of its length.

At the present time, there is practically no exact theory known which can be said to express the extension or contraction of various materials, nor even of different samples of the same kind of material, because the physical conditions of the several pieces are not exactly alike; hardness or softness and other conditions, due to previous treatment of the constituents of the specimen, step in to modify the result. The present object is, to show to the student the practical aspect of the case, and to explain the properties of some common materials, as they are met with in the workshop; but in the latter part of the book, when considering the principles which determine the form and arrangement to be given to structures, it will often become necessary to fix arbitrary rules, founded on the average result of numerous experiments, and in the application of which the judgment will have to be exercised.

That property by which bodies tend to occupy a determinate bulk, or in the case of solid bodies a determinate bulk and figure, under given pressures and at a given temperature, is termed their *elasticity*. By virtue of this property, they oppose a resistance to forces tending to change their volume or figure; and, if they have been deformed by the application of such forces, they return more or less completely to their original bulk and figure when those forces cease to act. A body is said to be perfectly elastic, which returns exactly to its original bulk, or, if a solid, to its original bulk and figure, after such distortion or straining, when again placed in the same conditions of pressure and temperature as at first. A body which imperfectly returns to its original condition after straining is said to be imperfectly elastic. For all solid bodies, there are known limits to the amount of straining force which can be applied, without producing a definite and measurable change of figure or *permanent set*. The limit of straining force which can be applied, without producing any measurable permanent set, is termed the limit of perfect elasticity or, more simply, the limit of elasticity for the given material under the given kind of straining force. For less straining forces the body is, for practical purposes, sensibly perfect in its elasticity. For greater straining forces it is sensibly imperfect in its elasticity. It is found by experiment that, up to the limit of elasticity, the displacements suffered by the molecules of the body are sensibly proportional to the stresses which cause them, so that a double displacement is caused by a double straining force; a triple displacement by a triple straining force; and so on.

This property of elasticity is the more perplexing, because the deformations of solid bodies are so minute, and often so difficult to detect or to measure. Wrought iron is usually considered as one of the most elastic metals within certain limits, still its elasticity is far from perfect, except with a very slight stress, and most probably it may yet be found to be imperfect, even with the smallest load. Nevertheless, for practical purposes, Hooke's law, which supposes all bodies to be elastic within certain limits, may be accepted as sufficiently near to the truth.

The elasticity of materials, when considered as a practical question and as affecting the mechanical arts, is an important physical property. It is a property which is largely taken advantage of for most purposes, and for some other purposes it has to be carefully counteracted or guarded against.

The perfect elasticity of some solids within certain limits of straining force, is proved by this circumstance, that many such bodies suffer an innumerable number of repetitions of straining action, without being sensibly altered. A steel watchspring will work on for a century and give no marked symptom of change, and a twisted steel or even iron rod will shut a door for an unlimited number of times, and a chain or rod of iron will lift weights and submit to be lengthened and shortened constantly, for a long period, and apparently will recover its normal shape when the force is removed. But, most probably, this is so only because our means of observation are incapable of detecting the change which is gradually taking place.

Those other materials, which do not exhibit this property to the same extent, or which readily retain any new form which may be imparted to them, are usually considered as non-elastic, and indeed justly, for practical purposes, although, strictly speaking, they are only very imperfectly elastic, as is evident when they are subjected to careful experiment: such materials are often termed *plastic* materials, as for instance, soft lead, putty, clay, and similar substances.

Aëriform bodies, such as steam, air, and other gases, have no elasticity of figure, but are perfectly elastic as regards volume, for they may be expanded or compressed to an extraordinary extent, the expansion or contraction under constant temperature being sensibly proportional to the force employed to act upon them.

A very common error exists respecting the elasticity of liquids, such as water : they are frequently treated as if they were without elasticity ; but this is not strictly the case, except as regards elasticity of figure. Water does not seem to admit of being drawn out beyond the volume which is due to the pressure with which it is surrounded, and which in an open vessel is determined by the pressure of the atmosphere and its own depth. But when confined, water admits of a limited amount of compression, and when the pressure is removed, it instantly returns to its original volume, and so far is perfectly elastic. From the result of some careful experiments it was found that the pressure of one atmosphere reduced the bulk of water by nearly the $\frac{1}{20000}$th of its original volume, an amount so small as to be scarcely appreciable in practical operations, and which, as a rule, is disregarded. Even in the action of the hydrostatic press, working often with a pressure of 500 atmospheres, it may be neglected with no practical disadvantage; nor is the elasticity usually observable, in consequence of the greater movement of the surrounding metal interfering.

It is due to the great compressibility of elastic gases, such as air, and the incompressibility of liquids such as water, that machinery for working with the former can be driven at a much higher velocity than similar machinery working with the latter. From the greater compressibility of air or steam, and the work absorbed in compression, there is not that perceptible shock which occurs with water or with solids under similar circumstances. It is this compressibility which imparts to aëriform fluids such a degree of yielding softness when they are acted upon by mechanical apparatus.

The elasticity of solids is of a different character to that shown by gases or liquids, although many solids partake of the properties of the former in some degree. In solids, the elastic property is mostly shown in the more or less complete recovery of the original form, rather than in the recovery of volume, after compression or extension, and although within certain limits the elasticity of solids, such as the metals, may seem practically to be nearly perfect, still it is doubtful if they are entirely so. By very careful testing of long bars, it was found that they did not return to precisely the original length. This change is not easily observed in making experiments with short specimens, the alteration being so minute.

Another notable feature is the circumstance that the stretching of a rod, or chain, or bolt, when subjected to tensional strain, is not uniform unless the substance or strength of the bolt or chain is uniform. The greatest extension takes place at the weakest point, which is a great disadvantage; hence rods, bolts, or chains are frequently broken with less force, tear, and wear than would be inferred, if this fact were neglected.

The elastic property of solid materials takes a much more important part, in the stability of structures, than is usually apparent to the casual observer. A familiarity with the testing machine and the lessons which it teaches, cannot fail to render the constant presence of elasticity a living reality to the thinking mind, and this should never be lost sight of, because any variation in the length of parts that have to support each other introduces an element of weakness. The long bolt, or the long stay, may not, on this account, take a fair share of the work; and even in a single part, where the strain enters upon it by flanges, as in pillars, or by joints, as in tie-rods, unless we assume absolute perfection in the mechanical fitting, some portion may have more than its share of duty; hence it is that such end parts, so exposed, require to be made stronger than the main body.

Notwithstanding this, the elastic property, which is so prominent a feature in all the materials which are employed in the workshop, is of inestimable value, as without it the iron could not be hammered or otherwise roughly manipulated. The great freedom of treatment to which the metals may be subjected, in reference to their malleable and other properties, is greatly due to their elasticity.

When a rod of iron, or a beam, is in a normal or quiescent condition, its capacity for rigid duty is not equal to that of a similar rod or beam in which the elasticity has been partially used up, by bending it into such a curve as will not injure its future stability.

By using up, or calling into active exercise, the elasticity of a beam or other article, the strength, or rather the stiffness, is considerably increased, and great ingenuity is frequently shown by practical men in such adaptations, whereby a comparatively flexible beam, by being subjected to a bending force, is thereby enabled to afford an amount of rigidity considerably greater than would be obtained from a much stronger beam, when left in its natural state. A beam fixed between the walls of a building, on which are to be erected the supporting columns of a steam engine, will have a considerable amount of spring and vibration, unless the beam is inordinately strong ; but if this beam is firmly propped in the centre, and then bent down at the ends, by means of wedges within the wall boxes, the elasticity of the beam will be used up and the stiffness very greatly increased.

Another example is afforded in the case of long horizontal tie-rods. The elasticity of the rod permits it to bend into a curve, due to gravitation. But by first ascertaining how far the rod will bend, and bending it previously to the same extent before it is erected, and then fixing it in a position in which the camber is upwards, the after-bending will bring the rod down to a straight line. The rod under such conditions is not only more pleasing to the eye, but is in all respects more rigid, from the elasticity being absorbed by the arrangement.

In the familiar case of the springs of railway carriages, it is found advantageous to use up a considerable portion of their elasticity in the primary adjustment, by means of a rigid bar to which they are bound. By this means, the carriage does not sink as the passengers enter, and the springs are so arranged that the remaining elasticity comes into play when the carriage is nearly full.

It may appear inconsistent, to speak of a beam or rod being thus stronger by using up so much of its strength, but such is the case, for the class of purposes here indicated ; they are rendered stiffer with a given quantity of material, and it takes a greater stress to give further movement to the beam or rod.

In the application of wrought iron, steel, and all the

highly elastic materials, although they may be safely loaded up to near their limit of elasticity, still, owing to the uncertainty of perfection of the material, engineers seldom venture far beyond the half of the amount of stress, which the apparent limit of elasticity appears to warrant.

CHAPTER II.

ON THE EXPERIMENTAL TESTING OF MATERIALS.

BEFORE describing the nature and manipulation of a mechanical testing apparatus, it is necessary to make a few preliminary remarks, both in regard to the mode of carrying out experiments that are to be reliable, and to the care, precision, and accuracy which are required, and to draw attention to some of the many contingencies that may prevent the attainment of true results, or bias the mind of the operator, even where good instruments are employed, and with every desire to avoid error.

In making important experiments with solid materials, it is not advisable, and can never be satisfactory, to depend upon the result of an experiment upon a single specimen, however good it may appear to be, for the smallest defect or scratch upon the outside, or even a hidden fault in the interior, may modify its behaviour to an unknown degree. The stress which is applied will mostly expend itself upon the specimen at its weakest point, instead of being distributed throughout the mass operated upon. Three specimens at least should be tested, and the diameter, length, and quality of each specimen should be uniform, as far as may be practicable.

It is found, in the testing of short and long specimens, even of the same kind of material, that in general the short and long pieces do not all stretch alike, or in the proportion due to their respective lengths: the long bars stretch more than the short ones. The cause of this discrepancy is rather obscure, and it is doubtful whether it is owing to the long bar, from its mere length, having a greater proportionate risk of invisible defects, or to some other unknown cause; but practically it is found that greater uniformity is attainable with short specimens, possibly in consequence of the closer scrutiny to which they may be subjected. Hence it may be inferred, that in testing materials intended for a particular structure, the nearer the test bars approximate in their dimensions to that structure, the more reliable will be the average result for the particular duty. At the same time, the length does not affect the strength, for with the exception of weakness due to hidden defects, the actual ultimate tensile strength of any portion of a specimen is not much, if at all, affected by the length of the bar; each portion has to act independently and for itself, irrespective of the other parts, its behaviour being exactly according to the stress to which it is exposed, and this is not inconsistent with the disproportionate rate of extension of long and short specimens, referred to previously.

In testing bars or strips of metal, whether the pieces are long or short, but more especially in the latter case, it is of much importance that the stress should be applied to the specimen in a line coincident with the axis of the specimen; if it is not, the result will be erroneous, because, the stress not being uniformly distributed on the cross sections, one side will have to yield prematurely, and thus the resistance of the bar will be overcome in detail : for want of attention to this particular, many experiments made with rough apparatus do not afford reliable results.

Time is an important element in the movement of the molecules of the specimen. In carrying out experiments with metals in a testing machine, it is evident that the period of time during which the specimen is exposed to the stress must, as a rule, be limited. With a short duration of time and a short specimen no 'permanent set' can be

On the Experimental Testing of Materials. 11

detected, if the load does not exceed a certain limit. To that special limit attention is directed, as it is a most important one for the student to understand thoroughly; it is usually termed the *limit of elasticity*, and is as varied for different materials as the point of ultimate rupture. But the fact that the testing machine does not show permanent set until the apparent limit of elasticity has been exceeded, does not prove much, for it is quite possible that a long continued permanent load, or an often repeated application of a load not exceeding that limit, might ultimately fix the molecules in a new position; on this point, however, a difference of opinion exists, though the experience derived from thousands of carefully performed experiments, with short specimens, in an accurate machine, would lead to the conclusion, that within a certain range every metal returns to its original dimensions, as near as can be measured with moderately refined instruments.

It is more than probable, however, that the first application of a load, much less than that commonly supposed to mark the limit of elasticity, does produce a minute permanent set. From the experiments made by the Commissioners appointed to enquire into the Application of Iron to Railway Structures, it would appear that with the smallest stress applied, namely, ·56 ton per square inch, a minute, but perceptible, amount of permanent set was produced upon a bar 50 feet long. When the stress was equal to 1·69 ton, the set on the bar was equal to ·0025 inch, or the ·000,004th of the length of the bar, an amount so exceedingly small that it could not be measured in specimens only a few inches in length. For practical purposes, the permanent set in such cases may be disregarded, being inappreciable.

For the present purpose, it will be assumed that there is a limit of stress up to which every specimen is perfectly elastic, so far at least as our means of measurement permit us to ascertain. In making experiments on the value of different materials, one great object is to determine this limit of perfect elasticity. The operator must observe with extreme care the indication of the slightest permanent set, on the removal of the load. The limit of elasticity marks the maximum stress which can be exerted on a material without producing permanent deformation and therefore danger of ultimate rupture. But there still remains an amount of strength beyond the elastic limit, and which is a margin of safety in reserve. If a bar has been strained beyond the elastic limit and has taken a permanent set, it is, paradoxical as it may appear, in one sense stronger than it was before; that is to say, it will require a greater strain to be applied, in order to move the molecules a still greater distance and to cause them to take an additional set. This is an important fact, and it is at once observable, when making experiments with short specimens, that no new load which may be applied, less than that which produced a given permanent set, will practically carry the effect farther onward. Therefore, by taking a permanent set, the limit of elasticity has in fact been raised, as the bar will not permanently stretch any more, without being subjected to a still greater strain. Even this, however, is not strictly correct, if the loads are much in excess of that which corresponds to the elastic limit.

The elasticity of the several materials that are employed in the workshop is extremely unequal, both in degree and in the manner in which the condition is shown. For example, some solids have elasticity combined with hardness and brittleness in a high degree, and are similar in character to glass; such are some kinds of cast iron, hard steel, and certain mixtures of copper and tin, in which, tested by impact, the elasticity seems as perfect as in an ivory ball, yet, in the testing machine, the extent of movement is so small that it can scarcely be measured by fine instruments, and for any practical purpose cannot be taken advantage of; while, on the other hand, there are some of the metals that

On the Experimental Testing of Materials.

seem to have the properties of india-rubber, but in a less degree; such is wrought iron, in a certain state, and still more so mild cast steel, when tempered in a particular manner, which developes the property of elastic flexibility in an extraordinary degree.

In preparing various classes of specimens for the testing machine, the strength of the machine itself must be first taken into account, and this will determine the size of the specimens according to their strength. It is usual to have the several specimens carefully prepared in a lathe, of sizes nearly inversely as their strength, or in some proportion to their respective tenacity. It is extremely convenient, and it simplifies the subsequent calculation, to make them of such a diameter that their sectional area will be a convenient multiple or fraction of a square inch; say, for instance, one square inch, half a square inch, or one quarter of a square inch. The Figures 1, 2, and 3, show the form and relative

FIG. 1.

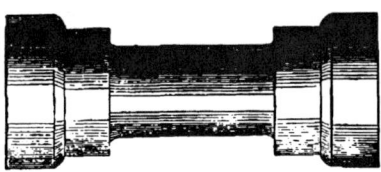

FIG. 2.

FIG. 3.

sizes of such specimens of cast iron, wrought iron, and steel, as are commonly used in the testing machine now to be described, and which has been used in the Royal Arsenal at Woolwich, since 1854. The drawing is made to a scale of one-half of the usual size of the actual specimens, when prepared for the testing machine, and the peculiar shape which is given to these three specimens is that which is used for experiments on tensile resistance, the middle portion being the part which is to be tested, and which is carefully turned in a lathe, so as to be perfectly parallel throughout; the ends of the specimens are enlarged as shown, in order to enable the machine to take a firm direct hold, and the advantage derived from the round form is chiefly this, that the specimens are produced in a lathe, with extreme accuracy, at a comparatively small cost for labour, and are in every respect as convenient and serviceable as if they were square, or of any other form.

Specimens for ascertaining the resistance to a compressive stress are generally made in the form of short solid cylinders, of such dimensions as can be overcome by the power of the testing machine, and therefore are so simple in form that a further description of them is unnecessary.

In the construction of the earlier mechanical testing apparatus, the mechanism generally consisted of a simple lever, which was mounted upon knife-edges at the several centres of motion and suspension, the short end of the lever laying hold of the specimen by a suitable bridle, and the weights being applied at the other end of the lever in the same manner as in a weighing beam. Great accuracy might be arrived at with such a simple lever, and some modern machines are fitted upon this principle and supplied with every requisite for testing the strength of the weaker class of materials, such as bricks, stones, mortar, Portland cement, glue, the bite of nails in wood, or for similar purposes; but they are not so convenient as those on the compound-lever arrangement, when great straining forces are required.

In the modern construction of testing machines, intended to operate upon very large or very strong specimens of different kinds of metal, the lever arrangement is being superseded by some modification of the hydraulic press. The hydraulic press affords the most convenient means of giving the necessary strain, but in the older machines, the means of measuring the strain applied were imperfect. The best kind of hydraulic machines are so contrived that the precise force, which is exerted by the water, is shown by a delicately adjusted steel-yard, or some other modification of the lever principle. The lever is selected as more certain and reliable than pressure gauges, because, however carefully the latter may be constructed, they are liable to alteration and error. In addition, when the load has to be calculated from the pressure in the press cylinder, the friction of the ram must be allowed for, and this cannot be done with any great accuracy.

CHAPTER III.

ON A MACHINE FOR TESTING THE STRENGTH AND ELASTICITY OF MATERIALS.

THE apparatus now to be described is intended to show the strength and elasticity of materials in various ways. It is probably one of the best and most correct machines which has yet been made, for the purpose of testing small short specimens, and for affording extreme accuracy in the results, so far as is possible with short specimens. The machine is shown in Figs. 4 and 5. This form of testing machine originated in the United States of America; the first one was brought to England from that country in 1854 by the author : since then many similar machines have been made, variously modified in size and arrangement, and have

16   *On the Strength of Materials.*

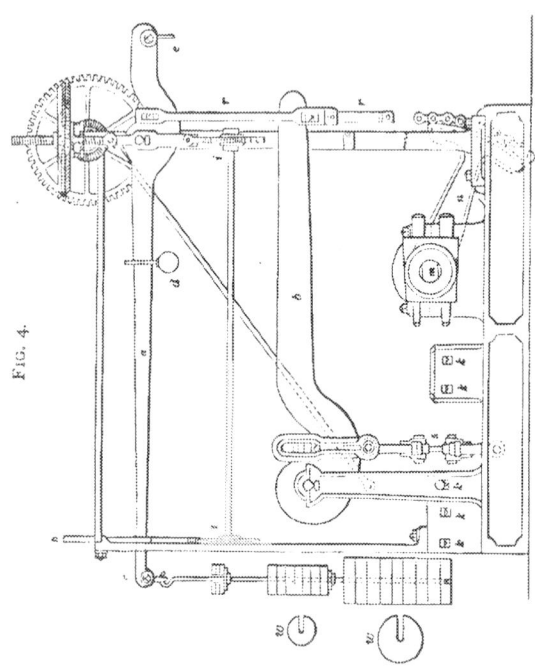

FIG. 4.

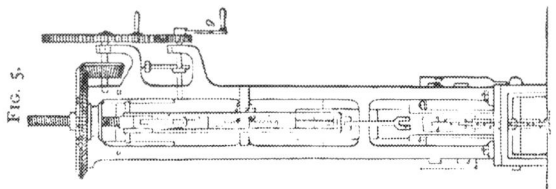

FIG. 5.

found their way to other countries. The machine in daily use at Woolwich, with which, during the last fifteen years, many thousand experiments have been made is also of this construction; it is principally employed in testing specimens to ascertain their tensile and compressive resistances.

The extensive application of machines on this principle is chiefly due to their simplicity and compactness of construction, and the great convenience which their several arrangements afford for various classes of experiments, as well as their extreme accuracy. They are provided with suitable bridles, holders, and other apparatus, to test tensile, compressive, transverse, and torsional resistances, and are adapted for experiments on the force required to punch or shear, on the hardness or softness of bodies, and on the flow of solids.

Figure 4 is a side and Fig. 5 an end elevation of this machine; the former shows most of the details, and will enable the general arrangements of the machine to be understood, with a little explanation. It consists of a combination of two levers, a and b, which together give a purchase of 200 to 1; that is to say, 1 lb. applied to the end of the upper lever at c will exert a stress of 200 lbs. on the specimen at s, and as all the bearing points of the entire lever apparatus are hard knife-edges, on hard smooth surfaces, the friction is reduced to a minimum.

Where so much accuracy is aimed at, it will be necessary, before commencing an experiment, for the attendant to see that the machine is nicely balanced, not only in regard to its own members, but likewise with reference to the various appliances which may have to be employed in order to carry out the experiments, any of which may disturb the equilibrium. This adjustment of the balance is effected by the small weight d on the upper lever, which is used in the steel-yard fashion, until the machine is accurately adjusted; if the want of balance is the other way, the adjust ment is made at the opposite end of the lever at e, by the suspension of as much weight as is requisite to secure perfect adjustment and to render the action of the machine as delicate as that of a weighing-machine beam.

The testing weights used are, for convenience, of a peculiar form, to adapt them for being placed upon shelves on the small rod of iron, which is shown suspended from the end of the lever at c, each weight being accurately adjusted, so as to exert a definite stress upon the specimen under operation. The plan of the weights is shown at $w\ w$; they are round in form, with a slit extending from the outside to the middle, to enable the operator to slip them easily into position upon the suspension rod. Several sizes of weights are used: the largest weighs 25 lbs., giving a strain of 5,000 lbs. on the specimen—of these there are nine; of the second size there are ten, each weighing 5 lbs., or giving a strain of 1,000 lbs. on the specimen; of the third size there are also ten, each weighing $\frac{1}{2}$ lb., or giving a strain of 100 lbs. on the specimen; thus making a total of 56,000 lbs., or 25 tons, which is the greatest stress that can be safely exerted by the machine.

The operator is, by past experience, enabled to judge of the effect which will be produced on the specimen, by the respective weights as they are applied one after the other, and so is able to load the specimen gradually up to the required limit. As the critical point is being approached, he uses smaller sized weights, until they are equal in effect to a larger one; he then removes the smaller weights and puts on a larger one as their equivalent, and continues with the smaller size, until their aggregate weight is again equal to that of a larger, and so on until the end is attained. When extreme accuracy is necessary, the utmost care has to be observed in the application of the weights, so as to avoid all rashness in the mode of carrying out the experiments.

When a specimen is subjected to a strain, it immediately

commences to stretch, and as the leverage of the machine is 200 to 1, will the stretch be magnified in the same proportion; that is to say, a stretch of $\frac{1}{100}$th of an inch in the specimen will cause the end of the lever to drop and the weights to sink 2 inches. Such a condition would be inconvenient, and therefore has to be provided for, and in this machine the effect of the stretching is compensated by the arrangements of the machine, which admits of adjustment at the point of suspension, the fulcrum of the upper lever being raised or lowered by means of a screw. As the specimen stretches, the fulcrum is proportionately raised, and this raising of the fulcrum is continued simultaneously with the stretching, and therefore the upper lever is constantly being raised and kept in a horizontal position during the operation. As it requires considerable power to manipulate the screw, a train of bevil-wheels and spur-gear is employed as an auxiliary; this part is seen more clearly in the end elevation at g. When the screw is running down, or when the strain is not great, the spur-pinion is dispensed with, and a handle is slipped into the spur-wheel, then the pinion is thrown out of gear, and thus the operation of tracing the limit of elasticity is greatly facilitated, by the promptitude with which the change in position of the levers may be effected.

If there were no provision for the jerk, with which the upper lever would necessarily go down, when final rupture takes place, it would be liable to be broken or bent at the termination of each experiment; this is prevented by means of a sliding stopper h, which is made of wood and is constructed with a slit or opening through which the end of the lever is passed, with just sufficient room to give the lever freedom to play, and in order to have this sliding board always in a proper position of readiness for the fall of the lever, this sliding stopper is moved up or down by the same screw movement which raises the fulcrum. Corresponding racks are placed at each end or side of the machine,

20 *On the Strength of Materials.*

with a horizontal spindle or shaft, geared at *i i*; hence the movements of both the fulcrum and board are simultaneous,

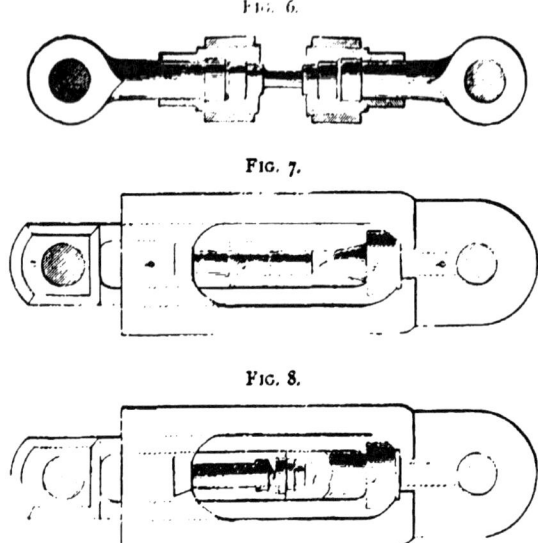

FIG. 6.

FIG. 7.

FIG. 8.

and the lever is caught by the wooden stop, before it has fallen a quarter of an inch.

In the Figs. 4, 5, the testing machine is shown as it would be arranged for ascertaining the tensile properties of materials. By referring to Figs. 1, 2, 3, the form of specimen will be seen; and in Fig. 6 is shown an enlarged view of the holder for tensile experiments. The ends of the specimen are held in carefully made split sockets, that fit the ends exactly, the two halves of the socket being kept together by rings or collars, which are slipped over and embrace them firmly without any effort or adjustment, and which may be easily removed by simply tapping them with a wooden mallet. Thus the change from one specimen to another is made without difficulty.

The machine being ready for an experiment, and the specimen in place, we may suppose that a weight is applied. If the specimen is of wrought iron, and is subjected to a stress of one ton per square inch, the middle or parallel portion of the specimen will perceptibly elongate, a distance equal to the $\frac{1}{10000}$th part of its length, and the end of the upper lever will sink 200 times that amount. If a proper measuring gauge is pushed between the shoulders of the specimen and accurately applied, it will be shown that such is the case, and, so long as the stress is continued, it will remain thus stretched; but if the strain is now removed, it will in time return again to the original length, thus showing that the material is so far elastic. On reapplying the stress, it will again stretch, and with every addition of one ton it will take an increment of extension in about the same proportion, and again and again return, if the weights are removed, nearly to its original place. This will continue up to a certain limit of load, of about 10 tons per square inch for wrought iron; under that point the specimen will return almost to its former dimensions, but beyond it the return will not be nearly so perfect.

It has to be explained, however, that the statement in the preceding paragraph, that each ton of stress applied to a square inch of wrought iron will cause an elongation of $\frac{1}{10000}$th of the length, is only approximately correct. This amount of elongation is rather over than under the usual amount of extension; but it is so near that, for all practical purposes, it may be accepted as correct. For each additional ton of strain, the bar will stretch another $\frac{1}{10000}$th, until the limit of elasticity is reached, which as a rule is found between 8 and 12 tons, according to the quality of the wrought iron tested. Hence, 10 tons is usually considered to be the average limit of elasticity of moderately good wrought iron, and the total stretch, up to that point and with that load, amounts to very nearly $\frac{1}{1000}$th of the length of the part operated upon. It is nevertheless very difficult to ascertain the true limit of elasticity, and published results often show great discrepancies as to the limit at which permanent set was first observed. In such cases, a judgment must be formed as to the value of the results, which depends on the accuracy of the testing machine employed, and the care and skill of the experimenter.

Some very careful experiments were carried out by the Commissioners, who were appointed to enquire into the Application of Iron to Railway Structures, with especial reference to this point, and the results are embodied in a Table at page 58. If such a course of experiments were repeated a few hundred times, so as to confirm the result, the natural law might probably be inferred, and, as will be seen by a reference to the facts in the Table, the minute results therein shown are consistent with the more general or approximate data given above. Still, that valuable Table only serves for the one or two specimen bars which were experimented upon; and those who have made many experiments are always impressed with the extreme variation of almost all the properties of materials, in different specimens, and especially of the limit of elasticity and the point where permanent set commences. In some specimens which were cut transversely from a large mass, the elastic limit was found to be under four tons of strain per square inch, while in other specimens of the same iron, but in the form of good rolled bar, of smaller sizes, it was found to have risen to about 12 tons per square inch, and generally that particular iron had a limit of elasticity ranging from 8 to 12 tons.

When the elastic limit is reached by the operator, then the future behaviour of the bar under trial will altogether depend on the precise nature of the iron. If it is soft and ductile, the iron will be drawn out to a much smaller diameter in the neighbourhood of the point of fracture, before the final rupture takes place. Although under such conditions it is usual to consider the breaking load as so much per square inch, calculated from the original dimensions, still, in point of fact, the ultimate strength is really more than that; because, from the altered diameter of the specimen at the moment of fracture, its area may have been reduced to $\frac{3}{4}$ths of the original area. This peculiarity is sometimes termed toughness; such iron will afford good warning before breaking, and is consequently preferred for purposes where repeated tension has to be exerted.

During the performance of the foregoing class of experiments, the operator has to watch carefully the behaviour of the specimen, in order to note its general character, for by continuing to increase the weight, gradually, upon the end of the lever, the whole of the characteristics of the specimen develope themselves, more or less clearly, and the appearances observed will much depend on its own inner nature. In some, the metal will flow, or be drawn in the heart of the bar only, thus leaving a corrugated exterior surface from the crumpling of the outer skin; in other specimens the flow is more uniform, and the outside is comparatively smooth. If, on the other hand, it is hard and rigid, it may not be drawn out to any great extent, but may break, with very little reduction of sectional area, and exhibit a high tenacity. If, on the contrary, it is of a soft and fluent nature, it will flow freely and be drawn out to a considerably smaller section, and then will break at the point where the diameter is most reduced. It may even now give a total strength varying from 20 to 25 tons per square inch of the original dimensions. The testing machine is equally suited for any other kind of metal; and, in dealing with familiar materials, such as cast iron, steel, copper, bronze, or other alloys, in order to arrive at their tensile properties, the same course is pursued as with wrought iron.

In arranging the machine to test compressive resistances, the shackles which hold the specimens for tenacity are removed, and another description of instrument is put in the same position in the machine. This instrument is shown in Fig. 7. It consists of two parts, *a* and *b*, the one sliding within the other, one of the parts being attached to the lever, and the other part to the framing of the machine. The specimen for this purpose is in the form of a small cylinder; weights are applied to the end of the upper lever, producing a stress 200 times as great on the specimen, in consequence of the leverage, which is the agency employed in compression. As movement takes place, either from the elasticity or the permanent set of the material, the fulcrum of the lever has to be moved, so as to keep the proper position; this is accomplished by turning the handle, which gives motion to the vertical screw, shown at *g* in Fig. 4.

A specimen-holder, nearly similar to that used for compression, is likewise used for other purposes, such as punching, shearing, or indenting, and for testing the hardness or softness of materials. Such a specimen-holder is shown at Fig. 8. It is here arranged for testing the force required to produce a certain amount of indentation, and is applied to the machine in the same manner as the specimen-holder for compression.

In Fig. 4, in the side elevation of the testing machine, are shown a row of points marked *k, k, k, k, k*—these are knife-edges firmly secured, and are used in testing transverse resistances. In Fig. 9 is exhibited one of the usual arrangements of the apparatus, when employed to test the transverse strength of materials. The bar, or rod, or small girder, is held up against two knife-edges, *k, k,* and the load is applied at the centre; the points of support are some definite distance apart, which is easily measured in this case. The distance shown is 10 inches; but by looking to the points *k* in the machine, it will be seen that provision is made for increasing the distance to 20 inches or 30 inches. The bar is kept up to these knife-edges by the knife-edge contained in the holder at *l*, which is so constructed as to embrace the bar, freely, during the experiment. In this instance, as in all other cases, the weights

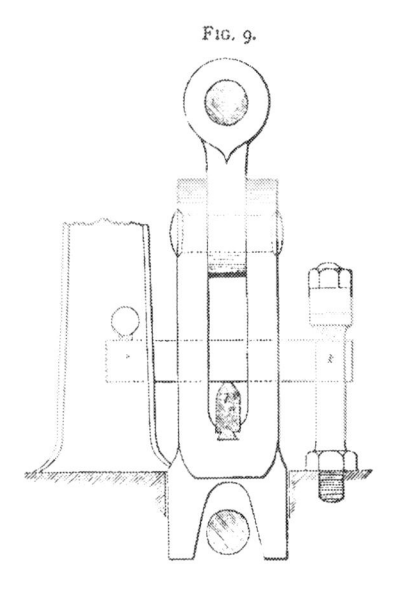

FIG. 9.

are applied gradually to the end of the lever; at the same time the behaviour of the specimen is observed, in regard to its strength, elasticity, buckling, set, &c. Experiment shows that the strength of rectangular bars, supported at the ends and loaded at the centre, is inversely as the distance between the supports, and directly as the width or thickness of the specimen, and as the square of the depth. The width of the specimen is that dimension which is perpendicular to the plane of flexure, and the depth is the dimension in that plane.

When a specimen is loaded transversely, it immediately

commences to bend in a curve, which, in the case of wrought iron or soft steel, from the change of form in the cross section, indicates considerable movement to have taken place, amongst the molecules composing the part most affected. The respective parts of the bar under tension and compression seem to meet, or run into each other, but not in a line that would be indicated by a previous knowledge of the tenacity or compressibility; this is an element that should be taken into account in any calculation of the strength of structures, when built up of the flowing metals.

Again, looking to Figs. 4 and 5, at the point m, there is shown the end of a specimen which is secured to the frame of the machine by means of two cotters; and the object of this arrangement is to ascertain the resistance of the specimen to torsion, or to a twisting strain, like that which is developed in the case of a shaft employed for conveying motive power. Any form of bar may be secured by the cotters. The bar is fixed securely at one end, or it may be at both ends, but by having a vacant space between the fixings, sufficient room is left for a lever to be firmly secured in the middle of the specimen. The position of this torsion lever is shown at n; and, as will be seen, the lever terminates with the outer end formed into a segment of a circle of some considerable range. To the lowest point of the segment, there is fixed a suitable pitch-chain, which is carried round the segment and upwards, to be hooked on to the lower end of the suspension rod r, by which it is connected with the lever a, in the same manner as for tension and compression.

It will be observed that the balance of the levers may be a good deal disturbed by this apparatus; the machine has therefore to be adjusted by the weights d or e, after which the testing may be proceeded with.

The point most necessary to be determined is the limit of torsional elasticity, which will be referred to in the Chapter relating to Torsion. Experiments with bars of different sizes show that the torsional strengths of shafts of different sizes are to each other nearly as the cubes of their diameters, any departure from that ratio being probably due to some accidental cause.

In making experiments, it is instructive to observe the deformation of ductile materials such as wrought iron and the softer steels, and to consider the action of the molecules composing the specimen, both when under tension and compression. The change of form which is observed can only be readily understood, by considering the metal as a fluid, the iron behaving in a manner similar to that of water passing through a tube or channel of any form. When a bar is drawn out, the principal flow of the, apparently, solid metal, is in the middle of the stream; and hence the peculiar sectional form which is assumed either by a round or square bar, or one of any other shape, showing that the farther the molecules of the material are removed from the centre of the flowing current, so much the less are they affected by the influence of the general movement. This unequal flowing of the molecules, may partly account for the apparent weakness of thin plates as compared with round bars of the same sectional area.

With a flowing, malleable, or ductile metal, the round bar when under tension is drawn out to a small diameter, uniformly, all round, but the metal goes in the middle chiefly, and the outside is shrivelled; while, with a rectangular or square bar, the flat surfaces are slightly hollowed, to an extent proportionate to their distances from the centre of the flow. Thus the corners become more prominent than they previously were. When cast iron is treated in the same manner there is no perceptible change of form, volume, or specific gravity, whereas with the flowing metals both volume and specific gravity are altered, the volume being increased and the specific gravity diminished. The strength is in some small measure interfered with by changes of form to which the material has been subjected during its manufacture. Good fibrous wrought iron is generally a little stronger in the direction of the fibre than transversely to the fibre, but the difference in any case is very small, and the strength is commonly assumed to be practically the same in both directions. The behaviour of a piece of iron, in this respect, is quite different to that of a piece of wood, which is much stronger in the direction of the fibre than it is when fractured across the grain, or in any other direction. A bar of wrought iron when it leaves the rolls is in a condition of great restraint; the exterior is not in perfect equilibrium with the interior of the bar, which at the commencement of an experiment affects the elongation and permanent set. The first effect of the application of a load is to liberate the constrained surface, and true conditions on which to form an opinion do not exist until equilibrium is established in the bar itself. Previous to that the result is deceptive; hence the advantage of carefully turned specimens.

The structure of a rope or a bundle of fine wires does not accurately represent the condition of materials, such as wrought iron, or even wood. It will lead to a false conclusion, if we reason on the assumption that, in materials like wrought iron, the fibres of which it appears to be composed are detached and independent of each other, or that they slide with freedom as in a rope or bundle of wires. On the contrary, they are firmly joined together, side by side, with a force nearly, if not altogether, equal to their general tenacity. The idea of a flowing stream, in which the velocity of the different parts depends on the stress applied and on their mutual adhesion and friction, represents much more truly the condition of a ductile bar under strain.

It is very interesting and instructive to take a square bar of iron, the larger the better, and carefully bend it round a large mandrel, into a circle, and then to observe the alteration of sectional form which ensues, the thinning away of the outer side and the increase of thickness at the interior, and the curved lines that gradually form and shade away from the one side or corner to the other. We can scarcely realise the changes that have taken place in the interior of such a bar, but they are strange and wonderful, and instructively illustrate many of the foregoing remarks, on the behaviour of a malleable, ductile, or flowing metal.

The testing machine here described may appear complicated, but it is really a simple machine ; any appearance of complication arises from the smallness of the diagram, or an imperfect appreciation of its mechanism and manipulation. It is easily used, little besides care and patience is required to arrive at accurate results.

By means of the various contrivances shown in the diagrams, combined with other expedients, almost any of the physical properties of materials may be ascertained by the testing machine, with sufficient accuracy for the guidance of those who have to apply them in the operations of daily life. Next to an acquaintance with the natural laws of mechanics, and to being familiar with the contrivances that have been devised by men in all ages, for turning the natural laws to account, a clear perception of the several properties of the materials that are to be employed in construction will be found most useful, indeed the value of such knowledge can scarcely be over-estimated.

CHAPTER IV.

CAST IRON.

THE usefulness of the different materials by which we are surrounded, measured by the extent of their application in the mechanical arts, has varied greatly in course of time. In past ages wood occupied a much more prominent position as a material of construction than it does now, having been superseded by metal for many of those purposes for which it was, formerly, exclusively employed. This substitution of metal for wood would even have been still more complete, if the former possessed some of the peculiar properties which render the latter extremely valuable for certain purposes. Amongst the metals, iron—either as cast iron, wrought iron, or steel—occupies now the foremost place among the materials at the disposal of the engineer. It is not strange, therefore, that, from their great importance in the arts, iron and steel should have been the subject of more experimental research than has been bestowed upon any other material. At the same time our knowledge is far from perfect, for there yet remain many obscure points, of great importance, which require further and frequently repeated investigation.

Cast iron is the crude metal derived from the smelting furnace. The ore and fuel are thrown into the furnace together, an intense heat is generated by means of a strong blast of air, the refractory ore is thereby reduced, and the iron gradually melts and runs down to the bottom by gravity.

Iron ore is very refractory, and in general is found mixed with earthy materials. Hence, it cannot be reduced by the carbon of the fuel alone, but requires the addition of fluxes, capable of combining with the earthy materials of the ore and of facilitating their fusion. If the ore is argillaceous, or contains clay, then the flux employed has to be of a calcareous nature ; whereas, if the ore is calcareous then clay is required as a flux, or, what comes to the same thing, the two sorts of ore may be mixed in suitable proportions, so that the one acts as a flux to the other. When the flux or third material is required, it is thrown into the furnace along with the ore and fuel, and at a high temperature it unites with the earthy matter of the ore and becomes slag, setting the greater part of the iron free.

It will thus be seen that, at the very threshold of the iron manufacture, there are several causes in operation which may seriously affect the quality, as well as the cost, of the iron produced. The liquid iron having to be in such intimate contact with the fuel and flux and their impurities, its quality is necessarily exposed to danger and may be materially affected by contamination with sulphur, phosphorus, or other injurious substances, present along with it in the smelting furnace.

During this preliminary smelting process, the cast or liquid iron has ample opportunity of combining with and absorbing a considerable quantity of carbon; this absorption of carbon in cast iron, whether in combination with the iron or not, is its distinguishing feature and determines its behaviour in most respects. It is the presence of carbon which gives to it its fusibility and enables it to be remelted again and again, and thus renders it suitable for the founder, the degree of fusibility depending on the quantity of carbon which it contains.

The presence of carbon renders the iron more liquid when in the fluid state, and softer and tougher when in the solid state. Still, when the iron has an excess of carbon, it is not so strong as iron with a less proportion of carbon ; hence the practical knowledge and judgment of the founder requires to be exercised, in the employment of the different sorts and in mixing those of different qualities, in order to obtain a metal with the requisite hardness, softness, closeness of grain, strength, and toughness, for the different kinds of castings which he has to produce.

When iron contains carbon in great excess, a portion of it may be in an uncombined condition, and this influences the quality in the direction of fluidity, softness, and weakness. The proportion of carbon in cast iron varies from 5 per cent. to 2 per cent. Cast iron may be poured into moulds of any form, and when carefully treated has considerable tenacity and even toughness and compressibility. But at the best it is comparatively an uncertain metal, and gives little, or, indeed, no warning previous to ultimate fracture, which is a radical defect.

In the preparation of cast-iron guns, where great strains are to be resisted, the conditions to be aimed at are rather contradictory. It might be inferred that the strongest sorts of iron would be the best for this purpose; but guns made of such iron fail at proof from their brittleness, and a softer mixture stands the proof much better. On the other hand, the interior surface of the bore when made of soft iron is not sufficiently close in the grain. The opposite conditions thus indicated to be desirable, can only be obtained by a compromise, the iron used being hard enough not to be spongy, and soft enough not to be brittle.

The best results, both in regard to elasticity and strength, are obtained by mixing a number of different kinds of cast iron, all carefully selected. Such a combination gives a higher result at the testing machine than the average of the different samples, when cast separately, yet the strength of the mass has seldom an ultimate tenacity exceeding 9 tons per square inch, and much oftener it is found nearer to 8 tons. Sometimes the tenacity reaches 14 tons, and cast iron has been produced with a tensile resistance of 15 tons, but such a tenacity is rarely attained. The average ultimate tenacity of ordinary cast iron is about 7 tons, and in inferior qualities only 5 tons, or even less, but these are exceptional qualities, and are of no value for purposes where strength is of consequence. The quality used for guns should have a tenacity not less than 10 tons per square inch.

In submitting cast iron to the testing machine, its limit of

Cast Iron. 33

elasticity, as shown by short specimens of common quality, is found to be rather low, or about one-third of its ultimate tenacity, but rather over than under; hence, in works of construction, it is not considered safe to strain ordinary cast iron, in tension, above 2 tons to the square inch, and even with the higher qualities, the greatest working tensile stress should never exceed 3 tons. For structures exposed to impact, the limiting stress should be much less, say 1 ton. It must be clearly understood, however, that these figures are only approximate. The student should closely study the Tables given at page 39, containing experiments on cast-iron bars of 10 feet in length, from which it would appear that the limit of elasticity is much less than we have assumed above, being under $\frac{1}{10}$th of the ultimate strength, and in which a permanent set was produced with a stress of $\frac{3}{4}$ of a ton; probably it would have been detected earlier, if the rod had been 100 feet instead of 10 feet long. Still, for practical purposes, it would be inconvenient to be governed by such minutiæ of measurement, and the approximate figures given in Table I. are sufficiently accurate for common purposes.

The following Table gives the extensions and elasticity of good cast iron, when the ultimate strength is about 10 tons per square inch of sectional area, which is considerably above the average for cast iron of commerce; as before stated, it may be considered as the lowest quality admissible for guns.

The first column shows the weights applied in lbs.; the second column gives the load on the bar per square inch of its section, in tons; the third column gives the visible measurable stretching while the stress is acting, within one minute of the application of the last increment of load; the fourth column gives the permanent set when the weight is removed, or shortly afterwards; the fifth column shows what may be considered to be the average elastic extension of a good specimen of cast iron.

By looking down the columns it will be seen that the

elasticity is apparently perfect with 4 tons, but that with 5·2 tons there is a perceptible permanent set, equal to the quarter of a thousandth part of an inch, and although the elasticity continues to the end of the experiment, still the limit of perfect elasticity has been reached at some unknown point between 4 and 5 tons, or well up to half the ultimate strength, which is higher in proportion than with the common cast iron of commerce.

Table showing the properties of cast iron, of the lowest quality suitable for guns:—

TABLE I.

Weight applied in lbs.	Stress in tons per square inch of section.	Visible stretch under the stress at end of half minute.	Permanent set when the load was removed.	Difference between the 3rd and 4th columns, showing the elastic extension.
4000	2·0	·0005	—	·0005
6000	3·0	·001	—	·001
7000	3·5	·0015	--	·0015
8000	4·0	·002	—	·002
10400	5·2	·0025	·00025	·00225
11600	5·8	·003	·0005	·0025
12500	6·25	·0035	·0075	·00275
14000	7·0	·004	·001	·003
14800	7·4	·0045	·00125	·00325
16000	8·0	·0055	·00175	·00375
17100	8·55	·007	·0025	·0045
17900	8·95	·008	·003	·005
18800	9·4	·01	·0045	·0055
19600	9·8	—	·0155	—
20500	10·25	—	·014	Broke

From the circumstance that the ultimate tenacity or tensile strength of castings of cast iron may vary between 15 tons and 5 tons per square inch of section, much attention has been given to the subject, both on the part of makers and of purchasers of castings, in order to secure qualities suitable for particular duties—more especially has this been the case where strength alone was the chief consideration. It may be said that, for the majority of ornamental castings, strength is of less importance, as for them the chief consideration is that the metal employed shall be extremely fluid, so as to enable it to flow like water, and fill up every ramification of the mould. In a less degree, the makers of the great majority of castings, for lathes or machine tools, aim rather at securing closeness of structure with a moderate degree of hardness than great strength, because mass for its own sake is of value in such articles, and this involves the presence of a quantity of material which renders the use of strong cast iron of less importance.

For beams and girders, strength and rigidity are the first considerations. Some founders, who are very careful in regard to the strength and practical goodness of their cast iron for such purposes, go to the trouble and expense of first melting the several sorts of iron, collected to form the intended mixture, and running the mixture into pigs. These are afterwards broken into fragments, which are examined one by one and carefully selected ; those pieces which present the best fracture are laid aside for important castings, the other pieces being kept for castings of less importance. It is also usual, with such careful founders, to cast test bars, in order to ascertain, in a rough and ready manner, the approximate strength of the metal they use for their own guidance.

A common method is to cast in a dry mould a bar 1 inch square by 54 inches in length, which is laid upon supports 48 inches apart, and weights are suspended in the middle until it deflects $\frac{3}{4}$ of an inch. Some sorts of cast iron are found too rigid to deflect much ; but good tough iron will do so invariably, and return again apparently uninjured. From hundreds of experiments made as above, it was found that a good mixture, properly cast, will sustain a load of 620 lbs. with a deflection of half an inch, while some have gone down the same distance with 240 lbs. Founders have ascertained that some sorts of cast iron, which are comparatively weak by themselves, will have their strength greatly increased by being mixed with another sort of iron. Hæmatite iron, for example, sustained, as above arranged, 480 lbs.; yet, when mixed with some Welsh iron, not of a stronger character, it carried a load of above 500 lbs.; thus showing that the question of mixtures of cast iron is an important subject in itself, which opens a wide field for the scientific founder.

Other founders, who contract for castings, are less careful, and have to be checked; hence, some engineers in their contracts for castings insist upon specimen bars being cast and tested, both for deflection and tensile strength. Such bars, of a section 2 inches by 1 inch, with a distance of 3 feet between the supports, are required to sustain a load, in the middle, of 30 cwts., and to deflect, before fracture takes place, at least ·29 of an inch. A bar 1 inch square is required to sustain a tensile stress of $11\frac{1}{2}$ tons per square inch, which is a very high tenacity. Such a course is highly to be commended, and helps to stem the downward course to the cheap and worthless; the student will do well to note the above, and he will find that, in the long run, it is the course which will best answer his purpose—weak and cheap material may do for a time, but its employment invariably brings its own punishment.

It will be evident, that the mere casting of test bars does not afford absolute security that the castings shall be of the same quality. Some engineers, contracting for a large number of beams or girders, on which the stability of an important structure has to depend, have experienced considerable difficulty in obtaining definite security that the proper quality of iron has been employed by the founder. In the case of cast-iron beams of the usual form, namely, with a vertical rib, and having the upper and lower flange in the proper inverse proportion to the tensile and compressive strength of the iron, they bend each girder separately by hydraulic pressure until they severally deflect the $\frac{1}{480}$th part of their length. The pressure applied, as shown by the water

Cast Iron. 37

gauge, is said to be about the half of the force required to break them; such a ratio, however, must depend on many conditions. A safer course, which is resorted to by some, is to cast an extra beam to every score that are required, or some other proportionate number, and then to test such a proportion of the beams cast, selected at random from the whole number, until actual rupture ensues; if the test beams break under a given load, the whole number are rejected.

Tables II. and III. refer to cast iron, and are compiled from valuable experiments made by the Commissioners appointed to enquire into the Application of Iron to Railway Structures. The first Table has reference to the behaviour of the bars under tension, and the second Table to their behaviour under compression, the bars being 1 inch square and 10 feet long.

From the length of these bars, and from the care bestowed upon the experiments, certain important data are furnished by them, especially in regard to the elongation and compression, and the accompanying permanent set, with loads for which they are inappreciable in short specimens. The Tables would have been still more perfect, if the temperature had been noted.

The actual measure of the permanent stability of any material is the point at which permanent elongation commences, and that point should have the chief attention. Still, if we are to trust these experiments, some judgment will be required, because, as Table No. II. shows, this point of commencing permanent set was only equal to 1-10th of the ultimate stress required to produce fracture, it is therefore necessary to take into account the ultimate strength as well, in order to know the margin of strength that lies beyond the elastic limits. With short cast specimens, this elastic limit, if in any degree observable, would appear to stand at a much higher point than that shown by these experiments, and, as before stated, it is usually considered to lie between the third and the half of the ultimate strength.

As this Table shows, the iron is in some measure elastic even to the end, and yet can scarcely be said to be perfectly elastic even at the beginning of the experiment. Only a few experiments were made with long bars. If a similar course could be gone through with every description of cast iron—hard, soft, weak, and strong, in all their varieties—engineers would then be furnished with more clearly defined knowledge on these points than exists at present. Such an enquiry should be made, with special regard to the relative fitness of different sorts of cast iron for different purposes. Precise data thus obtained would be of great advantage to practical men.

From several hundred experiments made at Woolwich with selected specimens of the higher qualities of cast iron, the ultimate tenacity was found to range from 10,866 lbs. up to 31,480 lbs., the average tenacity being 21,173 lbs. per square inch. This average result, although high, is under 10 tons, and lower than the average of 850 samples, sent in for competition, the tenacity of which ranged from 9,417 lbs. up to 34,279 lbs. per square inch.

It will be observed that the best of these specimens had a tenacity over 15 tons, and the worst a little over 4 tons per square inch. Similar experiments, carried out on a number of specimens of the ordinary cast irons of commerce, gave an average of 12,912 lbs., or a little over 6 tons, and some specimens of Nova Scotia iron gave an average of 15,821 lbs., or a little over 7 tons per square inch. The foregoing were the ultimate breaking strains, and the specimens were only 2 inches in length. They gave no warning of fracture, that was perceptible to the senses with ordinary instruments; but no doubt there was movement, if it could have been measured by suitable means, with verniers, for instance, that would read to the 100,000th part of an inch.

Table No. II. shows the elongations and amount of permanent set with a given load of cast-iron bars one inch square and 10 feet long; these bars are of the same

Cast Iron. 39

size as the wrought-iron bars of Table VII. A comparison of these Tables will show a great difference, irrespective of strength, between the behaviour of the cast and the wrought iron; the relation of weight to extension is not nearly so uniform in the cast iron.

Table No. II. also shows that a permanent set took place with a load of ·7 of a ton, or less than $\frac{1}{10}$th of the breaking stress, which is not in accordance with the general notion; the set, however, is so small that it could not be measured in a short specimen. The sixth column is the most instructive, it shows the relation of weight to extension, which is far from being constant.

Synopsis of experiments on the extension of bars of cast iron, 1 inch square and 10 feet long:—

TABLE II.

Weights in		Extensions in		Sets in	Ratio of weight to extension $\frac{w}{e}$
lbs. = w	tons.	Fractions of an inch = e.	Fractional parts of entire length.	fractions of an inch.	
1053·77	·47	·0090	$\frac{1}{13333}$	set not visible.	117086
1580·65	·70	·0137	$\frac{1}{8758}$	·00022	115131
2107·54	·94	·0186	$\frac{1}{6451}$	·000545	113309
3161·31	1·41	·0287	$\frac{1}{4181}$	·00107	110150
4215·08	1·88	·0391	$\frac{1}{3069}$	·00175	107803
5268·85	2·35	·0500	$\frac{1}{2400}$	·00265	105377
6322·62	2·82	·0613	$\frac{1}{1957}$	·00372	103142
7376·39	3·29	·0734	$\frac{1}{1634}$	·00517	100496
8430·16	3·76	·0859	$\frac{1}{1398}$	·00664	98139
9483·94	4·23	·0995	$\frac{1}{1207}$	·00844	95316
10537·71	4·70	·1136	$\frac{1}{1058}$	·01062	92762
11591·48	5·17	·1283	$\frac{1}{903}$	·01306	90347
12646·25	5·64	·1448	$\frac{1}{828}$	·01609	87329
13699·83	6·11	·1668	$\frac{1}{718}$	·02097	82133
14793·10	6·60	·1859	$\frac{1}{645}$	·02410	79576
16664·00	7·43	mean breaking weight.			

Apparent Limit of elasticity, $\frac{1}{10}$ of breaking weight.

Although the property of tenacity is that which receives

the greatest attention, for many practical purposes the resistance which is offered to compression is also of great importance.

Cast iron offers a considerably greater resistance to compression than it does to extension, and the same may be said of the other metals. In a series of experiments made with a variety of metals, to test their properties in this respect, the several kinds did not differ so much from each other as might have been expected; the metals employed were cast iron, wrought iron, and steel, and the specimens were in the form of short cylinders. The object for which the experiments were made was, to ascertain the force required to shorten a specimen, permanently, to the extent of $\frac{3}{1000}$th of an inch in the direction of the length. To produce this amount of shortening required a stress which was found to range between 30,500 lbs. and 40,700 lbs. per square inch of area, the length of the specimen operated upon, in each case, being 1 inch and the diameter ·533 inch. This was not the force required to crush the specimen into a cake, or into fragments, which is not the point which it is of greatest practical importance to ascertain. For practical purposes, we require to know the pressure at which the surface of the specimen begins to yield or give way, and this pressure or stress is termed the elastic limit in compression; *cæteris paribus*, that is the best material which requires the greatest pressure to produce the result.

Of ten specimens, cut from cast-iron guns of high quality, the softest was found to yield with 30,000 lbs., while the hardest specimen was able to sustain 40,300 lbs., and the average resistance of the ten specimens was about 35,000 lbs. Of ten specimens of wrought iron, cut from forgings of high quality, the softest began to yield with 22,800 lbs., and the hardest with 31,000 lbs., the average being 26,900 lbs. In each case weight was added until the specimen became shorter, by the $\frac{3}{1000}$th of an inch.

Table No. III. gives the compression and 'sets' with 10

Cast Iron.

feet bars of cast iron; the bars were prevented from bending or buckling, by means of a case or mould in which they were contained.

Experiments on compression of cast-iron bars 1 inch square and 10 feet in length:—

TABLE III.

Weights		Compressions		Sets in fractional parts of an inch.	Ratio of weight to compression $\frac{w}{c}$
in lbs. = w.	in tons.	Fractions of an inch = c.	Fractional parts of the length.		
2064·74	·9217	·01875	$\frac{1}{8304}$	·00047	110120
4129·49	1·84	·03878	$\frac{1}{3091}$	·00226	106485
6194·24	2·76	·05978	$\frac{1}{2007}$	·004	103617
8258·98	3·68	·07879	$\frac{1}{1523}$	·00645	104823
10323·73	4·60	·09944	$\frac{1}{1206}$	·00847	103819
12388·48	5·52	·12030	$\frac{1}{997}$	·010875	102980
14453·22	6·44	·14163	$\frac{1}{847}$	·01405	102049
16517·97	7·36	·16338	$\frac{1}{734}$	·01712	101102
18582·71	8·28	·18505	$\frac{1}{648}$	·02051	100420
20647·46	9·21	·20624	$\frac{1}{581}$	·02484	100114
24776·95	11·04	·24961	$\frac{1}{480}$	·0322	99263
28906·45	12·88	·29699	$\frac{1}{404}$	·043	97331
33030·80	14·72	·35341	$\frac{1}{338}$	·06096	93463
37159·65	16·56	·41149	$\frac{1}{291}$	·08421	90304
Bar much undulated					

Limit of elasticity, $\frac{1}{17}$ of weight which permanently injured the bar.

Table No. IV. is also from the Report of the Commissioners; and from the Tables here given, together with the other results of experiments referred to in this chapter, the nature of cast iron may be approximately inferred, at least so far as regards its more prominent characteristics.

42   *On the Strength of Materials.*

TABLE IV.
Synopsis of Experiments on Extension of Cast Iron.
Form: Round rods or bars 50 feet long. 1 square inch area of section.

Name of Iron.	Number of Experiments.	Weights in lbs.	Weights in tons.	Extensions in fractions of an inch.	Extensions in fractions of entire length	Sets in fractions of an inch.	Mean breaking weight per square inch.	Mean ultimate extension.
Lowmoor, No. 2	2	2117 6352 10586 14821	·94 2·83 4·72 6·61	·0950 ·3115 ·5740 ·9147	$\frac{1}{6318}$ $\frac{1}{1924}$ $\frac{1}{1045}$ $\frac{1}{655}$	·00345 ·025 ·06425 ·12775	16408 lbs. = 7·325 tons.	1·085 inch, or $\frac{1}{553}$ of the length.
Blaenavon, No. 2	2	2096 6289 10482 13627	·94 2·87 4·67 6·83	·09422 ·3065 ·5770 ·8370	$\frac{1}{6367}$ $\frac{1}{1957}$ $\frac{1}{1040}$ $\frac{1}{718}$	·00268 ·01675 ·0575 ·11475	14675 lbs. = 6·551 tons.	·0325 inch, or $\frac{1}{873}$ of the length.
Gartsherrie, No. 3	2	2109 6328 10547 14766 15820	·94 2·82 4·78 6·59 7·06	·09225 ·3177 ·5802 ·9452 1·0487	$\frac{1}{6505}$ $\frac{1}{1848}$ $\frac{1}{1035}$ $\frac{1}{634}$ $\frac{1}{573}$	·001 ·0145 ·0475 ·11325 ·13812	16951 lbs. = 7·567 tons.	1·167 inch, or $\frac{1}{514}$ of the length.
Mixture, No. 3 Leeswood No. 3 Glengarnock	3	2107 6322 10536 12643	·94 2·82 4·70 5·64	·0914 ·2667 ·5349 ·6702	$\frac{1}{6576}$ $\frac{1}{2252}$ $\frac{1}{1123}$ $\frac{1}{895}$	·00376 ·01823 ·04321 ·06417	14812 lbs. = 6·6125 tons.	·8095 inch, or $\frac{1}{741}$ of the length.

Cast Iron.

The following Table refers to experiments, made at Woolwich, with short cylinders of soft cast iron under compression:—

TABLE V.

Length of specimen, 1 inch; diameter, ·533 of an inch.

Weight applied per square inch into tons.	Compression in decimals of an inch.		Elasticity as shown by difference between the visible and permanent compression.
	Visible.	Permanent.	
3·8	·003	·0005	·0025
11·0	·004	·001	·003
17·6	·006	·002	·004
20·6	·007	·0025	·0045
50·0	crushed		

The stress required to produce a complete crushing of the specimens is shown by the two following Tables:—

TABLE VI.

Experiments on the crushing strength of Cylinders of Cast Iron, made by Eaton Hodgkinson, Esq., for the Commissioners.

Description of Iron.	Diameter of specimen in inches.	Height of specimen in inches.	Mean crushing weight per square inch of section in tons.	No. of Experiments.
16 various sorts	·75	·75	39·37	48
	·75	1·5	38·28	48

By reducing a number of other experiments to a common form, we obtain the following results:—

On the Strength of Materials.

Abstract of various Experiments on the Crushing Strength of Short Cylinders of Cast Iron.

Authority	Nature of Iron.	Mean crushing weight weight per square inch.
Hodgkinson for Commissioners	From various parts of the kingdom	38·8
Fairbairn . .	Carron, hot blast, No. 2 . .	54·35
Do. . .	Carron, cold blast, No. 2 . .	55·975
Do. . .	No. 1 Iron, from various parts of the kingdom . . . . . No. 2, do. do. do. No. 3, do. do. do.	40·118 42·68 58·675 } mean 47·16

The following abstract, from the researches of Sir William Fairbairn, C.E., Bart., shows the gain or loss of strength from remelting cast iron :—

'Abstract of experiments on the crushing strength of the same iron, after successive remeltings:

'Decreased in strength from 44 to 40·17 tons, by remelting four times, then gradually increased from 40·17 at the fourth remelting to 95·9 tons at the fourteenth remelting.'

'Abstract of experiments on the transverse strength of the same iron, after successive remeltings:

'Increased in strength and elasticity up to the twelfth remelting, and then gradually decreased in both properties from the twelfth remelting until, after being remelted eighteen times, it only possessed ⅔ths of its original strength, and the ultimate deflection of the bars had decreased from 1·44 to ·476 inch.'

The following Table contains results of experiments, on the transverse strength of cast iron, at various temperatures, by Sir W. Fairbairn. It may be useful to note the 11th experiment, such exceptional results are constantly met with in practice :—

TABLE VII.

These results are reduced to 1 inch square, and 2 feet 3 inches between the supports.

Nature of Iron.	Temperature.	Breaking weight in lbs.	Relative power of resisting impact.
Cold blast No. 2.	27° 32 113 192 red in dark red by daylight	874·0 949·6 812·9 743·1 723·1 663·3	597·7 382·4 273·1 223·7 — —
Hot blast No. 2.	20° 32 84 136 187 188 red in dark	811·7 919·7 877·5 875·7 638·8 823·6 829·7	325·0 395·0 369·4 340·6 229·3 298·9 —
Hot blast No. 3.	212° 600	814·4 875·8	— —
Cold blast No. 3.	212° 600	924·5 1033·0	— —

These experiments teach us, first, that No. 2 cast iron is stronger at the freezing point to resist transverse strain and impact, than at any other temperature, and that when the temperature is raised from 32° to 113°, the cold blast iron loses 14·4 per cent. of its transverse strength, and 28·5 per cent. of its power to resist impact. Secondly, that the No. 3, or hard irons, are stronger at high temperatures than at lower ones, apparently the reverse of the result with the No. 2 iron; this, however, is probably owing to the increased ductility of these irons at the higher temperatures.

CHAPTER V.

WROUGHT IRON.

THE material most extensively used in the arts is the malleable, ductile, tough, fibrous, weldable material, usually termed malleable or wrought iron. This variety of iron, at the present time, is made directly from cast iron, by a process of elimination, whereby the greater portion of its carbon, as well as any sulphur, silicon, phosphorus, and other impurities are got rid of, as far as may be practicable, the change being effected by subjecting the iron, while in a hot or liquid state, to the oxidising influence of a powerful flame, by which the impurities, as well as the carbon, are carried away in the form of gas or combine with the slags in the furnace

This purification of cast iron gives the mass an entirely different nature and new characteristics. By the process, it becomes considerably stronger, it acquires a great degree of toughness, yet, unfortunately, it loses the capability of being cast in moulds. As a compensation, it acquires a new property, namely, the quality of assuming the viscous or sticky condition, so that when two or more pieces are brought together at the proper temperature, they may be united by the welding process, as it is termed, either by the blows of a hammer, by pressure, or otherwise.

When cast iron is thus acted upon by an oxidising flame, and every part is exposed to its influence in the puddling process, the newly-converted mass of viscous iron, when removed from the furnace, may be compared to a dirty iron sponge full of impurities, the doughy mass has to be put under a steam hammer, or some form of squeezer, in order to drive or wring out the mechanical impurities which it contains. It is then elongated, by means of rolls, into a rough bar and cut into short pieces; these pieces are piled up into a bundle, which is reheated to the welding point, and again rolled, so as to cleanse the iron thoroughly. Indeed, for the better descriptions of wrought iron, the processes of piling, reheating, and rolling are sometimes repeated several times, until the proper quality is attained.

The quality of wrought iron thus treated depends, to a great extent, on the original selection of the mixture of cast iron from which it has been made, as well as on the purity of the fuel used for the converting process. Iron has an inherent and great affinity for sulphur, phosphorus, or other impurities, which it has the opportunity of taking up; hence wrought iron of high quality can only be obtained by extreme care in every stage of its manufacture.

In the manufacture of those qualities of wrought iron in which purity is the most essential condition, mineral fuel is dispensed with, and charcoal made from wood is employed as a substitute. Such wrought iron is chiefly used for subsequent conversion into the better qualities of steel, by means of the process of cementation.

By whatever process the change from cast iron into wrought iron is effected, the decided alteration of the whole character for the better is unmistakable, and that the ultimate tensile strength should be increased to an average of 25 tons per square inch, and the limit of elasticity to 10 tons per square inch, by the elimination of carbon and other impurities, are very remarkable facts. Wrought iron is found to differ almost as much in quality as cast iron, and this is partly due to the fact, that it is seldom or ever entirely free from carbon or other ingredients. The amount of carbon varies from an imperceptibly small quantity up to ¼ per cent., wrought iron containing the latter proportion being almost equal to mild steel. The presence of a small quantity of carbon, while serving to increase the strength, rigidity, and hardness of the material, at the same time greatly interferes with the welding property; for this reason, it is much more difficult to weld the stronger or more steely kinds of wrought iron, than the softer, weaker, and less steely varieties. Pieces of soft iron go together, and unite into a homogeneous mass much more kindly than harder and less pure pieces, when raised to the welding temperature. The quality in this respect is easily ascertained, by making the iron red hot and plunging it into cold water, when the soft iron is found to retain its softness, but the hard iron becomes still harder, in a manner similar to the behaviour of steel, though in a less degree.

The production of large masses of cast iron, by melting the metal, is much more easily accomplished than the process of welding similar masses of wrought iron; in the former case the founder has simply to prepare an earthy or other refractory mould, to provide due means for its ventilation at the moment of casting, and then to pour out the fluid metal, which finds its level and fills up the empty space, in the same manner as water, when poured into a vessel. But, in the formation of large masses of wrought iron by forging or welding, the operations are of a slower and more expensive nature than those of the founder. The process of forging is a kind of gradual building up, bit by bit, variously conducted, to suit the individual circumstances of each particular case. In a general way, the operation of building up large masses of wrought iron, is accomplished by uniting an innumerable number of small pieces into blooms; then a number of blooms are united into slabs and smaller slabs into larger ones, until at length the ultimate dimensions of the required forging are attained. When it is borne in mind that a sound weld depends not

Wrought Iron.

only upon the purity and equal temperature of the surfaces, but likewise upon the absence of all vitrified oxide, dirt, or impurity of any description between the parts to be joined, it is manifest that the forging of immense masses of wrought iron is an operation surrounded with many practical difficulties, and its application, for purposes where perfect homogeneity of mass is essential, is therefore limited. It is, also, too often found that the excess of heating, in proportion to the hammering or working that can be given to the mass, is injurious, and that, consequently, the iron of the heavy forging is reduced in strength, when compared with the original iron, in the condition of rolled bar.

RESISTANCE OF WROUGHT IRON IN TENSION.

The ultimate tensile strength of wrought iron is usually set down as 25 tons per square inch, but this is above the present general average, 23 tons being nearer as an approximate round number. At the same time, hard steely wrought iron frequently has a tenacity of 30 tons, and inferior kinds of only 19 tons. As a rule, square or round bars are stronger than plates, by at least 3 tons to the square inch, but weaker iron is not inferior for some purposes; such iron frequently has the welding property in a marked degree, and is preferred in consequence, wherever welding has to be extensively resorted to, in converting wrought iron into the required articles.

Entire specimens, cut from a rough bar of wrought iron as it leaves the rolls, are generally found to be weaker than a portion of the same bar, which has been turned in a lathe before testing. If a portion of the same rough bar is hardened by any mechanical process, such as cold rolling, or swaging or hammering, the strength and hardness are both increased, but by annealing the specimen, the original conditions of strength and softness are fully restored.

In the process of rolling iron into bars or plates, the molecules and aggregates of molecules of the iron are elongated into what is usually termed fibre. It might, therefore, be expected that the bar or plate would be stronger when drawn asunder, with the tension in the direction of the fibre, than when drawn asunder, with the tension at right angles to the fibre. Such is generally the case, more especially when the mass of iron is thin as in boiler-plates, the difference sometimes equal to 20 per cent. Besides, the elongation of the fibre by strain is greater in the former direction than the latter, thus giving earlier warning of impending fracture.

The strength of wrought iron, as given in the older tables, ranges higher than would be the case if similar tables were now to be compiled, from specimens taken at random from the iron of commerce. Wrought iron has been quoted to have a tenacity of 34 tons, but unless the experiments giving that tenacity were inaccurately carried out, the iron must have been hard and steely; we still find wrought iron occasionally with a tenacity of 30 tons, and the writer has some specimens of 32 tons, but such iron is hard and almost unweldable, and is much more brittle than iron of a lower tenacity; hence, in selecting iron for any particular purpose, the peculiar strains to which it will be exposed must be taken into consideration, before determining the quality which should be employed.

The strength of wrought iron is not much affected by variations in temperature, when under 350° Fahr.; above that temperature, it begins to lose strength, and as it approaches to a dull red heat, the ductility greatly increases, and the flowing property comes into play and reduces the resistance fully one-half; hence the opportunity for the smith to 'strike while the iron is hot.'

There has been considerable difference of opinion, in regard to the strength of wrought iron when exposed to severe frost, but, from recent investigation, it would appear that practically it is not much, if in any appreciable degree, affected by the lowest temperature of an English winter. Nevertheless, there exists a popular notion that iron and steel are greatly affected by frost, and thereby rendered more brittle than when at an ordinary temperature, and this notion receives some support from the fact, that the fracture of rails and railway axles is most frequent in winter. But the fact is susceptible of another and more probable explanation, for in winter the roads are hard and rigid, causing great jar. The notion that the strength of iron is less in cold weather is not borne out by the general average of the experiments made in summer compared with the average of those made in winter; these experiments, however, are performed within a building.

That wrought iron is to some extent more ductile in warm weather, than in extremely cold weather, seems probable, even, although the tenacity is not much affected. This effect of the presence of heat may also, in some degree, justify the general belief which exists amongst practical men, of the greater liability to fracture in winter than in summer. It is frequently found, when guns are being proved to destruction by continued firing, that the first round in the morning, when recommencing the experiments with a cold gun, proves fatal to its endurance.

The appearance which is presented by a fracture of wrought iron depends greatly on the mode in which the rupture has been effected. If it is accomplished suddenly, the fibre is actually broken short, and the crystalline texture apparently predominates; whereas, if the fracture is produced by a slower process, the fibre of the bar is then made conspicuous, because the element of time is essential to enable the fibres to be drawn out from each other. All such appearances, however, are greatly modified by the quality of the bar which is operated upon; some judgment, as well as experience, being necessary to arrive at a just conclusion, and opinions formed from the fracture only are not always to be depended upon, even when the examination is made by an expert.

From many fractures of wrought-iron axles, chains, and other pieces, which have been in daily work for a number of years, the idea has become general, that with sufficient intensity of jar, repeated an indefinite number of times, a change takes place in the structure of the iron. This is not inconsistent with the results obtained by the testing machine. It is satisfactory to know, however, that such a result may in a great measure be anticipated and prevented by simply subjecting the axle, chain, or hook, from time to time, to the process of annealing, by which its original condition is practically restored.

At the same time, it may be stated that this is an obscure subject, for there is not much positive evidence of the fact, or explanation of the cause of such deterioration in wrought iron, after being exposed to long-continued working, as in the axles of railway carriages and in crane chains; still, if such is the case, it is now generally considered to be due to the effect of vibrations often repeated, when the part affected is for the moment loaded up to or beyond the elastic limit. For it has to be remembered that, in the case of the crane when surging, or in that of a carriage jolting, this element of motion suddenly checked produces a stress which has to be added to the ordinary stress existing when the several parts are in a state of rest. Cranes that are usually worked with less straining than that for which they were originally intended, seem to maintain their efficiency for an indefinite period, whereas other cranes, which are much employed up to their maximum ability, do sometimes give way unexpectedly, and with a less load than that to which they have been usually subjected. In some cases there is an excess of deflection on a particular part, which, being often repeated, has ultimately caused fracture.

Whatever may be the explanation, the phenomenon is now familiar to those who are engaged in practical operations, that when a piece of wrought iron is thus subjected to a long-continued series of blows, or violent jars, of sufficient intensity, or that call the full elasticity of the material into active exercise, rupture will sometimes take place prematurely, and must be expected sooner or later. The change to rigidity, which overtakes iron when worked cold, may partly account for some of the frequent fractures of the chains of cranes or other iron-work similarly exposed, and this view is in some measure supported by the fact that when such chains are annealed, at stated intervals, say annually, the liability to accident is greatly diminished.

Considerable difference of opinion and practice exists with regard to the safe load which may be put upon ordinary wrought iron; this of course must depend greatly on the nature of the structure, and the kind of stresses to which it is exposed. A crane, for example, with the risk of a load being suddenly checked, during lowering, by a rash attendant, must necessarily have a larger margin of safety than would be required for a structure with a steady load; for the latter, a strain of 5 tons might be ventured with safety, while for the former $2\frac{1}{2}$ tons would hardly afford the same measure of security; some judgment has therefore to be exercised in determining the strain that wrought iron may be entrusted with, but in no case should it be burdened with a load producing a strain of more than 5 tons per square inch of sectional area.

Wrought iron is frequently found to present peculiar phases of character, which must not be overlooked, as the conditions affect its strength; some descriptions of wrought iron, otherwise good, are found to be intractable and apparently brittle when red hot, yet are perfectly pliable when cold—this peculiarity is usually considered to be due to the presence of a small trace of sulphur in the iron. Then there are other descriptions of wrought iron having the opposite quality, being comparatively brittle when in the cold state, and perfectly pliable and workable when red hot. This quality is said to be due to the presence of silicon or phosphorus. These conditions are so marked as to suggest the idea that the gas of the impurity forms a coating, which envelopes or separates the molecules of the iron, and they greatly diminish the value and usefulness of the iron in which they exist, for many practical purposes. These defects are mostly due to the impurity of the original ore, and the ironworker usually corrects them in the course of manufacture, by combining ores of opposite nature in suitable proportions, so as to obtain an average quality; of these two defects the 'red shortness,' as it is commonly termed, is the less objectionable.

In wrought iron, and especially that of high quality, there is no fixed point at which permanent set begins to be observed; for although we speak of the limit of elasticity as somewhere in the vicinity of 28,000 lbs. per sq. in., or about half the ultimate tenacity, yet, at the same time, it has to be clearly realised by the student that such a statement is only approximate; and although round numbers may be easily remembered, and are sufficiently correct for most practical purposes, still they must be considered as approximate only, and taken with reference to the exact truth contained in the following Tables. The student is referred more especially to Table VIII., which throws more precise light on the facts than any general statements.

It has also to be observed that in making experiments, whether with long or short specimens, there is a marked difference between the behaviour of cast iron and wrought iron. It has been already stated that the former has comparatively little perceptible elongation, unless the bar is of considerable length, while the elongation of the latter can be seen distinctly, and increases even to the last moment. The final stress which is required to produce rupture must not be calculated from the original dimensions of the specimen. The final stress is the actual strength of the reduced area after stretching. This stretching is one of the distinguishing features of wrought iron as compared with cast iron, and is

Wrought Iron. 55

one of the special virtues of good wrought iron and mild steel.

When wrought iron is wire-drawn and its section reduced, the strength of the part so elongated is increased by the process. Thus, an iron wire, when made with iron of a strength of 25 tons, will have its tenacity increased to 35 tons per inch of sectional area by the mere process of drawing ; and the most remarkable feature of all is this, that the specific gravity is actually reduced at the same time, showing that the conditions which give strength are not solely dependent on the closeness of the molecules.

In the former part of this chapter attention has been drawn to different qualities or natures of wrought iron. When a variety of specimens of different kinds of wrought iron are under great tension, their respective behaviour will greatly depend on their own inherent hardness or softness ; a hard specimen will be found to elongate very little, and will ultimately fracture without reduction of diameter, while a softer specimen will be drawn out considerably, the middle part becoming gradually smaller, and fracture will ultimately take place at the smallest section, most probably at a lower strain than was the case with the harder iron. It might then be inferred, that the hard iron was the better of the two ; but such an inference might not be correct, as the latter would, practically considered, be much more reliable when subjected to jar or sudden strain. At the same time, in estimating the original strength of the elongated specimen, it will be objectionable to take the reduction of diameter into consideration because for practical purposes the elongation should never be called into exercise. The advantage of soft iron is, rather that it can be used with greater safety, than that a higher stress can be permitted where it is employed, for the softer iron is more likely to be drawn out than to be broken asunder, and gives ample warning before ultimate fracture.

In experiments made at Woolwich with ten short specimens, cut from heavy wrought-iron forgings, the average apparent limit of elasticity was only 23,760 lbs., while the average point of ultimate fracture by tension was equal to 48,160 lbs. The forgings from which the specimens were cut were all intended for gun purposes, and consequently were of high quality, or intended to be so. Similar experiments were made with rolled bars of the same high quality, both in the state of bars and when wound into spiral coils, and then subjected to the welding process, in order to form them into entire cylinders for gun manufacture. All the specimens were found to have suffered from forging, to the extent of 3,481 lbs. per square inch, on an average. The results were as follows :—

Limit of elasticity { Bar . . . . 31,000 lbs.
{ Cylinder . . . 27,852 ,,
Ultimate rupture { Bar . . . . 58,986 ,,
{ Cylinder . . . 55,500 ,,

No doubt, the loss of strength here shown was owing to the absence of sufficient working, to counteract the weakening tendency of excessive heating.

From some interesting experiments, that were made to ascertain the tenacity of the welds of different qualities of iron, and with differently formed welding surfaces of contact, it was found that with soft iron of the finest quality, heated in a clear fire, with scarfed joints, and both surfaces slightly rounded, the strength of the welds was equal to that of the original bar, or in round numbers about 25 tons per square inch; while with other descriptions of harder iron, variously selected, the strength of the welds was always less than that of the bars, the lowest and worst example, which was an exceedingly hard and steely specimen, having a tenacity at the weld of only 12,000 lbs. per square inch. A still lower average result was obtained in all attempts at butt-welding, with similar qualities of iron, even when the surfaces were rounded to give a clear way of escape to the vitrified oxide;

Wrought Iron. 57

the average ultimate tenacity was 32,140 lbs., or a little over half of the strength of the iron. With turned butt surfaces, when heated in a furnace flame, it was a little higher than the above.

Some experiments were made with bar iron of a very hard steely quality, its ultimate strength being about 30 tons, in the specimens cut from the bar. It was, of course, difficult to weld such iron, and impossible to secure butt welds with a tenacity greater than 10,000 lbs. per square inch, the steely property, due to the presence of carbon, being the barrier to a ready union of the pieces, when manipulated by the smith in the ordinary manner. It will thus be seen that the strength of a wrought-iron structure depends on many conditions, all of which have to be taken into consideration in estimating the reliance to be placed on it, or in calculating the requisite quality or quantity of material to be employed.

Table VIII. is a synopsis of the results of an experiment made to show the extension and permanent set of a rod of wrought iron 50 feet long, and which have been reduced in this Table so as to compare with the other experiments, of a similar nature, but made upon a rod of cast iron 10 feet long, as given in Tables II. and III.

Column No. 1 shows the relative value of the successive weights that were used in making the experiment, and is inserted for comparison with the relative value of the extensions that were produced by those weights, as shown in column No. 4; by means of this comparison will be seen the departure from a constant uniformity which might be expected.

Column No. 2 gives the actual weights in lbs. which were applied per square inch of section of the rod.

Column No. 3 gives the actual weights applied per square inch of section of the rod in tons.

Column No. 4 shows the relative value of the extensions produced by the successive weights, and teaches when

On the Strength of Materials.

TABLE VIII.

1. Relative value of the weights used, when the smallest is represented by unity.	2. Weight applied per square inch of section (in lbs)	3. (in tons)	4. Relative value of the extensions produced when the smallest is represented by unity.	5. Extensions On 10 feet of length in inches.	6. In fractional parts of the length.	7. Permanent sets on 10 feet of length in inches.	8. Ratio of the weight applied per square inch to the extension.
1	1261.8	.56	1.0	.0052	$\frac{1}{23074}$	perceptible	2426654
2	2523.7	1.12	2.2	.0115	$\frac{1}{10434}$	—	2194539
3	3785.6	1.69	3.2	.0169	$\frac{1}{7100}$	.0005	2239982
4	5047.4	2.25	4.3	.0224	$\frac{1}{5357}$	.0006	2253317
5	6309.3	2.81	5.3	.02772	$\frac{1}{4329}$	.0005	2276078
6	7571.1	3.38	6.3	.03298	$\frac{1}{3639}$	.00045	2295679
7	8833.0	3.94	7.2	.0379	$\frac{1}{3166}$	.0005	2330609
8	10094.8	4.50	8.2	.043	$\frac{1}{2790}$	.0005	2347642
9	11356.7	5.07	9.3	.04854	$\frac{1}{2472}$	—	2339662
10	12618.6	5.63	10.3	.0537	$\frac{1}{2234}$	.0007	2349828
11	13880.4	6.19	11.4	.0595	$\frac{1}{2016}$	—	2332847
12	15142.3	6.76	12.4	.0648	$\frac{1}{1851}$	—	2336775
13	16404.1	7.36	13.4	.0698	$\frac{1}{1719}$	—	2350165
14	17666.0	7.88	14.4	.0753	$\frac{1}{1593}$	.0013	2346084
15	18927.9	8.44	15.7	.0817	$\frac{1}{1468}$	—	2316752
16	20189.7	9.00	16.8	.0874	$\frac{1}{1371}$	.0027	2310038
17	21451.6	9.62	17.9	.0931	$\frac{1}{1288}$	—	2304145

Mean 2307600.

Wrought Iron.

18	22713·4	10·14	19·0	·0992	1/1105	·0041	228966·1
19	23975·3	10·70	20·3	·1057	1/1135	—	22682·4
20	25237·1	11·26	21·6	·1125	1/1034	·0068	224330·3
21	26499·0	11·82	23·1	·1204	1/571	—	220091·5
22	27760·8	12·38	24·7	·1288	1/951	·012	215534·7
23	29022·7	13·04	27·9	·145	1/817	—	200156·7
24	30284·6	13·52	38·3	·1991	1/107	—	—
—	—	13·52	38·6	·2007	1/67	—	—
—	—	—	—	after 5 minutes			
—	—	13·52	38·8	·2018	1/121	·0736	150894·8
—	—	—	—	after 10 minutes			
—	—	13·52	39·5	·2054	1/111	·0774	—
—	—	—	—	after 15 minutes			
—	same weight repeated	13·52	40·0	·208	1/170	·0796	—
—	,,	13·52	40·3	·2096	1/124	·0814	—
—	—	—	—	after 20 minutes			
—	—	—	—	·2366	1/160	—	—
—	—	—	—	after 1 hour			
—	same weight repeated	13·52	45·5	after bearing the weight 17 hours	1/107	·1082	—
25	31546·4	14·08	46·5	·242	1/136	·1083	130357·2
—	—	—	—	after 5 minutes			
—	same weight repeated	14·08	47·1	·2449	1/117	·1111	—
—	—	—	—	after 5 minutes			
26	32808·0	14·72	105·9	·5506	1/16	·4141	—
—	same weight repeated	14·72	135	·7024	1/11	·5635	46709
—	—	—	—	after 5 minutes			
—	,,	14·72	153·1	·7966	1/180	·6558	—
—	—	—	—	after 10 minutes			
—	,,	14·72	195	1·014	1/17	·866	—
—	—	—	—	after 15 minutes			

On the Strength of Materials.

1. Relative value of the weights used, when the smallest is represented by unity.	2. Weight applied per square inch of section (in lbs.)	2. Weight applied per square inch of section (in tons)	4. Relative value of the extensions produced when the smallest is represented by unity.	5. Extensions On 10 feet of length in inches.	6. In fractional parts of the length	7. Permanent sets on 10 feet of length in inches.	8. Ratio of the weight applied per square inch to the extension.
27	34070	15·20	258·8	1·346 after 1 minute	$\frac{1}{89}$	—	—
—	⸺	15·20	269·2	1·4 after 2 minutes	$\frac{1}{85}$	—	—
—	same weight repeated	15·20	307·7	1·6	$\frac{1}{75}$	1·44	21294
—	same weight repeated	15·20	317·3	1·65 after 1 minute	$\frac{1}{73}$	—	—
—	—	—	343·4	1·786 after 1 hour	$\frac{1}{67}$	1·628	—
28	35332	15·76	392·3	2·04 after 5 minutes	$\frac{1}{57}$	1·874	17320
—	same weight repeated	15·76	419·2	2·18 after 5 minutes	$\frac{1}{55}$	2·01	—
29	36593·8	15·76 16·33	433·4 488·4	2·254 2·54	$\frac{1}{53}$ $\frac{1}{47}$	2·08 —	14407
30	37856	16·90	556·5	2·894 after 6 minutes	$\frac{1}{41}$	loop broke	13081

compared with column No. 1—first, that the elasticity of wrought iron as here shown is not so perfect as is generally assumed, even when the strains to which it has been subjected are small. If the material had been perfectly elastic, the relative value of the extension would have corresponded, exactly, with the relative value of the weight applied, that is to say, twice the first weight applied would have produced twice the extension caused by that weight; three times the weight, three times the extension, and so on; but this is not the case. Second, that there is only a slight and gradual increase in the relative value of the extensions over that of the weight applied, until that weight exceeded twenty times the first weight applied, and when the material was subjected to a stress of 11·26 tons per square inch, or about one-half the ultimate strength. Third, that after this point is reached, the value of the extension produced exceeds the relative weights very rapidly, until when the rod is subjected to a stress of 16·33 tons per square inch for six minutes, it is stretched seventeen times more than it would have been if the material had been perfectly elastic.

Column No. 5 gives the actual extensions, and shows very clearly—first, that the length of time during which the material is subjected to a stress, has a considerable effect upon the result. To take an example from the Table, with a weight of 13·52 tons, or rather more than one-half the breaking weight of good wrought iron, the extension on 10 feet of length increased from ·1991 inch, when the weight was first applied, to ·2366 inch, after the weight had acted upon the rod for seventeen hours. Second, that by removing and replacing the same weight the extension is increased. (This is not observable to the same extent in testing short specimens.)

NOTE.—In each of the above cases it will be observed that the weight is more than one-half the ultimate breaking load.

Column No. 6 gives the extensions produced in fractional

parts of the length, and will be found to be useful for ascertaining the extension of wrought iron of any length by the application of any of the weights, per square inch, given in the Table.

Column No. 7 gives the amount of permanent set produced by the successive weights, and teaches—first, that with so small a weight as half a ton per square inch, a perceptible set was given to wrought iron; second, that the amount of set was very small, until the material had been strained up to one-half its ultimate strength, or about 11·26 tons per square inch, at which point the set was ·0068 or $\frac{2}{33}$rds of the extension; third, that the set rapidly increased after the stress had reached 11·26 tons per square inch, or one-half the ultimate strength, and when it reached 14·08 tons or about $\frac{5}{8}$ths of the ultimate strength, the set amounted to ·1083 inch or $\frac{1}{3}$ths of the extension, and when strained to 15·76 tons per square inch, or less than $\frac{3}{4}$ of the ultimate strength, the set was nearly $\frac{9}{10}$ths of the extension produced by that weight.

Column No. 8 only shows a comparative result, and the numbers there given are obtained, by finding the ratio of the weight applied in lbs. to the extension produced in inches, and had the numbers in this column been equal, then the material would have been perfectly elastic; but although not equal, yet the numbers in the first twenty observations are nearly so, and therefore show that the material was near to perfect elasticity under strains less than 11·26 tons per square inch; hence, we infer that the usual expression of 'the limit of elasticity' is admissible, and in the case of wrought iron is almost correct.

The mean of these twenty numbers is 230,760, and consequently the extension of this bar of wrought iron in inches may be taken at $\frac{1}{230760}$th of the weight applied in lbs. to stretch it, and the weight applied in lbs. to produce any given extension would have to be 230,760 times the extension in inches. Now the modulus of statical elasticity is generally understood to be the weight or force in lbs. which would stretch a bar to double its length, if its elasticity remained perfect, and as this Table shows the extension on a bar 120 inches long, we obtain the modulus when we find the weight required to stretch this bar 120 inches, namely, 230,760 × 120 = 27,691,200 lbs.

NOTE.—This number is variously stated in books upon the subject, and would be greater or less than that given above according as the material employed was more or less elastic.

It would have added to the value, and probably would have modified the results, if the expansion and contraction, due to the differences of temperature, had been taken into account, which does not appear to have been the case, judging from the description given in the Blue-book in which they are recorded.

RESISTANCE OF WROUGHT IRON IN COMPRESSION.

In the manifold structures for which wrought iron is now so extensively applied, the property of resisting compression is frequently called into active exercise. Whenever wrought iron is thus subjected to compression, in either large or small structures, a permanent deformation is produced, if the load exceeds a certain value. It may be seldom that the liability of this metal to give way by compression is observable, as compared with failure in tension, because the failure in the former case is more likely to occur from flexure or want of stiffness, which again may be due to want of fixedness. In structures subjected to thrust, much depends on the mode of attaching the several parts, such as struts or stays, to the adjoining parts. A great improvement has taken place of late years in this respect; the extensive employment of the lathe, planing-machine, and other refined tools, extensively used in modern workshops, permits the parts which have to be joined to be truly faced at the ends, so as to take the correct position firmly; and this can be done at such a moderate cost as to render it commercially practicable. The sound and uniform bearing at the connection, thus obtained, adds greatly to the strength of the whole structure. By secure fixing, the work to be done on the material, in order to cause it to deflect, is considerably more than doubled, without increasing the weight.

Referring to page 40, there are quoted certain experiments with wrought iron and cast iron as agents to resist compression, from which it will be seen that the stress necessary to shorten certain specimens of good wrought iron an amount equal to $\frac{3}{1000}$ths of an inch was, for the softest 22,800 lbs., the hardest 31,000 lbs., the average being 26,900 lbs.

The foregoing experiment was made with specimens cut from forgings. From experiments made with ten other specimens taken from rolled bar iron of high quality, the specimens having been reduced in a lathe from 3-inch bars, the softest specimen required 31,000 lbs., and the hardest 35,000 lbs., or an average of 33,000 lbs. It will be observed, that the iron of the bar was considerably superior to the iron cut from the forging, although the two were of similar quality originally, thus showing the effect of better working.

Table IX. contains the results of a number of experiments carried out at Woolwich with short cylinders of wrought iron under compression.

Structures scarcely ever fail practically from the actual crushing of the material; failure is more often due to the alteration of form which takes place, disturbing its fitness for the particular purpose for which it is intended. When the pillar, strut, or frame, is long, it generally yields by flexure rather than actual crushing. If properly stayed or trussed at intervals, a long rod will act as a pillar, taking the stress through the entire length, the elasticity being uniform throughout. But in order to save

TABLE IX.

Diameter of specimen, ·533 inch ; length, 1 inch.

Nature of Iron.	Weight applied in tons per square inch.	Compression in decimals of an inch. Visible.	Compression in decimals of an inch. Permanent.	Elasticity as shown by the difference between the visible and permanent compression.
Specimens cut from bar iron, varying from ¾ to 2½ inches square.	7·8	·0015	—	·0015
	11·6	·0025	—	·0025
	13·4	·0045	·003	·0015
	13·4	·0055	·0035	·002
	15·8	·0075	·0055	·002
	50·0	—	·245	—
	11·0	·002	—	—
	12·4	·003	·0005	·0025
	14·0	·0045	·002	·0025
	15·0	·0065	·0035	·0025
	16·0	·0075	·005	·0025
	50·0	—	·26	—
	12·0	·0015	—	·0015
	13·2	·0025	·0005	·002
	14·0	·004	·0015	·0025
	15·0	·006	·0035	·0025
	16·0	·008	·006	·002
	50·0	—	·245	—
	14·4	·0025	·005	·002
	14·8	—	·007	—
	50·0	—	·259	—
Marshall and Mill's Irons, soft and of fine quality.	11·8		·003	
	12·0		·0035	
	12·6	not noted	·004	
	13·2		·0045	
	50·0		·2655	
	14·4		·0005	
	15·0		·0025	
	15·6	not noted	·004	
	16·0		·0055	
	50·0		·2615	
	10·0		·0005	
	11·0		·0025	
	11·6		·0045	
	50·0		·286	

TABLE IX.—*continued*.

Nature of Iron.	Weight applied in tons per square inch.	Compression in decimals of an inch. Visible.	Compression in decimals of an inch. Permanent.	Elasticity as shown by the difference between the visible and permanent compression.
Taylor Brothers' Yorkshire Iron; in the direction of the fibre.	11·8 12·4 13·2 50·0	not noted	·002 ·0035 ·0055 ·2775	
	13·0 14·0 50·0		·015 ·004 ·287	
	14·0 14·5 50·0		·0015 ·004 ·249	
S. C. Iron. Specimen cut from a welded coil.	10·0 11·4 50·0	not noted	·001 ·004 ·318	
	10·0 11·2 50·0		·001 ·004 ·3105	
	11·8 12·8 50·0		·0005 ·004 ·288	
	12·0 12·6 50·0		·0005 ·0035 ·2845	
	12·6 14·2 14·8 50·0		·0005 ·003 ·0045 ·2565	

the expense and trouble of trussing or encasing long bars, when making experiments to ascertain the resistance to a

Wrought Iron.

crushing force, it is more convenient to deal with short specimens, otherwise deflection will commence before the molecules begin to flow.

By increasing the stress upon these short cylinders of wrought iron or soft steel, they are found to shorten gradually by bulging outwards in the middle. The effect of this change of form is to slightly stiffen the metal, and this affects the malleable or flowing property; unless the specimen is extremely soft, it will soon show symptoms of slight fissures or cracks at the part which is bulging. To prevent this, the annealing process must be resorted to, and with care the pillar can be flattened down to a thin disc, gradually presenting a larger surface for the machine to act upon. Reckoning the intensity of the ultimate pressure from the original dimensions, a stress of upwards of 100 tons per square inch is necessary to actually flatten down wrought iron.

When wrought iron or steel is flattened by compression, it might be supposed that the specific gravity would be increased; but such does not appear to be the case to any appreciable extent. The specimen either cracks and splinters, or finds relief by lateral yielding, the enlargement commencing in the middle of the cylinder.

CHAPTER VI.

STEEL

THE material termed steel is a nearly pure alloy of iron with a small portion of carbon. As a metal, it is closely allied both to cast iron and wrought iron, and may be made from either.

The most common mode of manufacture is to convert pure wrought iron into steel, by measuring out a definite portion of carbon for the iron to absorb; or, it may be made from cast iron by a reverse proceeding, namely, by the elimination of the carbon and impurities, allowing so much carbon to remain as will give it the required steely qualities.

Although steel thus forms a connecting link between wrought iron and cast iron, it differs greatly from the latter, cast iron being a crude, indefinite, and impure alloy, while steel is a purified alloy, containing a definite percentage of carbon.

Steels differ from each other in many respects, but more especially in regard to the degree of steeliness, and, within certain limits, any degree of steeliness may be obtained. In making fine steel by the cementation or common process, bars of pure wrought iron are enclosed in a fire-clay box, and surrounded with powdered charcoal. The whole mass is then kept at a red heat for a certain period; the porous molecular structure of the iron is opened, and the carbon vapour finds its way into the body of the iron. The bars of iron are subjected to this process for a period generally ranging from five days to a fortnight, or even longer, according to the quality required; the longer the process is continued, the more steely does the bar become.

Steel. 69

As some of the bars have a greater opportunity of absorbing carbon than others, and from other contingencies, the steeliness of the batch varies, and uniform quality can only be obtained by breaking the bars into small pieces, sorting these fragments into lots, judging by the appearance of the fracture, and then melting the lots in a crucible. The liquid steel is cast into an ingot, and hammered or rolled into the cast steel of commerce. In casting very large masses, a great number of crucibles are necessary, and as these must all be ready for pouring at the same time, and must be emptied into the mould consecutively, the successful casting of heavy ingots requires a very good organisation.

In the arts, cast steel is required of all degrees of steeliness. The milder sorts of steel are only a little more steely than the harder varieties of wrought iron, and the mildest quality may contain about ¼ per cent. by weight of carbon. The highest qualities contain as much as 1 per cent. of carbon, and there are many intermediate qualities. As compared with hard wrought iron, mild steel, while not containing much more carbon, is yet more perfectly homogeneous in its granular structure, and is superior to it both in strength, and in almost every other good quality.

A good serviceable quality of steel, for many purposes, is now extensively made by the 'Bessemer' process, which appears to be, at first sight, a reverse method to the cementation system described above. The Bessemer process commences with cast iron in the crude state, which is melted and poured into a vessel, and, while liquid, a strong current of air is passed through it. The carbon in the iron is burned out, by the oxygen contained in the air passing through the molten mass, and the other impurities are gradually eliminated, until at length the iron is in condition of comparative purity, and chemically similar to wrought iron. In order to make it into steel, there is added to this purified metal, a measured portion of a pure cast iron, generally that called '*spiegeleisen*,' or 'looking-glass iron,' containing a large and definite quantity of carbon, which converts the pure iron into a steel sufficiently good for an immense variety of applications. From the circumstance that steel may be produced by this process at a less cost than by the former method, it is, to a large extent, taking the place of wrought iron.

Steel, like wrought iron, is much improved in its nature by being thoroughly worked either under the hammer or by rolling. Like most metals, steel is found to be more or less porous after casting, which is due to small air or gas bubbles which have not been able to find their way to the surface; the effect of hammering or rolling is to consolidate the mass, and to render the grains of the metal finer, denser, and stronger.

During the last few years, Sir Joseph Whitworth has been engaged upon a course of most valuable although most expensive experiments, conducted with powerful apparatus. These experiments have for their object, the attainment of steel of great density and general goodness, by combining the best materials and the most thorough working, and using every care that the best skill can devise. In Sir Joseph Whitworth's system, after the liquid steel is poured into a metal mould of sufficient strength, a piston or plug is inserted, upon the top of which there is at once brought to bear the full force of 8000 tons of hydraulic pressure. The effect is at once perceptible, a corresponding pressure pervades the liquid steel, the porosity is overcome, and the metal shrinks rapidly by the sudden closing up of its pores. The shrinking continues for some time at a rate perceptible to the eye, and afterwards more slowly, for a period of nearly a quarter of an hour. Some of the shrinkage observed is due to the decrease of temperature, but it is mainly owing to the hydraulic pressure.

In considering the effect of pressure, as compared with a blow, in the consolidation of a mass of steel or iron, it would appear that the former should be the most effective, because the metal acted upon has not an opportunity of lateral yielding, whereas in either hammering or rolling, the metal flows away from the point of impact of the hammer, or squeeze of the rolls. Such is the view taken by Sir J. Whitworth. Hence when the ingot, which has been compressed in the liquid state, is taken out of the mould, it is brought to a working temperature, and is then subjected to a series of squeezing operations by hydraulic pressure, squeeze after squeeze being applied until the required dimensions are arrived at; this hydraulic pressure does not act without control, a dial and pointer shows the amount of compression of the mass, and so prevents any risk of over squeezing and making the article too small. The whole process is carried out with such ease, and is so gentle in its action upon the tools that are employed, that on seeing the process performed, it is impossible not to feel that it is a great step in the right direction, and it is encouraging to find a man with the courage to go into such gigantic enterprises, for the purpose of advancing practical science.

The goodness of steel may be said to depend on three things: first, the materials selected; second, the nature of the working to which it is subjected; and, third, the care taken by the makers. These three conditions apply to all descriptions of steel, from mild to high, and each quality is equally good for its own special applications.

The quality of fine cast steel suitable for cutting tools, contains a larger percentage of carbon than either the milder varieties of cast steel, or the varieties of Bessemer steel, which are taking the place of wrought iron, or even the fine cast steel that is used for the lining of guns. The gun steel contains about ·03 per cent. of carbon, and in its natural state has an ultimate tenacity of from 30 to 35 tons, but when made red hot and cooled in oil, its ultimate strength rises to 40 or 45 tons per square inch. The apparent elastic limit of short specimens rises in an equally remarkable degree, being for untempered steel specimens 13 to 15 tons, and for steel tempered in oil, 28 to 32 tons per square inch. The toughness is also increased, so that the tempered steel may be bent, twisted, or drawn out to a degree, far beyond that to which it would submit in the untempered condition.

It will thus be seen that the limit of elasticity of mild cast steel, when tempered in oil, is fully equal to three times that of wrought iron. This is an important consideration, which will in time determine the extensive use of that material, and, more especially, because it stretches so much before final rupture takes place, which is a very valuable and important property.

The marked increase in the tenacity of cast or Bessemer steel, as compared with wrought iron, has already led to the application of the latter for many purposes where iron was formerly used, and wherever strength requires to be combined with lightness. In the case of rough structures, such as girders or bridges, the use of steel will gradually advance, the chief difficulty in the way of its more general application being, not so much the difference of cost, as the fact that the engineer is unable to determine whether good steel has been really employed by the contractor; testing would settle the point, but the trouble and expense of testing every bar or plate is a serious practical barrier to the adoption of this plan. It has been suggested that the specific gravity of a cutting from each part of a structure might be taken, steel being 0·1 per cent. heavier than wrought iron, but the difference is so small that the distinction of the quality by that test is rather too delicate for practical purposes. It has also been proposed to take some of the punchings from the plates, to draw them out to a small bar at a smith's forge, and to test them either for tenacity, or by tempering with fire and water in the usual manner. All such testing, however, is not in accordance with the usual practical notions of the workshop at the present stage of our progress.

In a wide range of experiments made with ordinary cast steel, when in its natural or untempered state, the ultimate tenacity was found to vary from 114,000 lbs. down to 67,000 lbs., the highest being a little over 50 tons, and the lowest a little under 30 tons per square inch; but specimens of steel are often met with both of greater and of less strength than the foregoing. A cast-steel specimen of extreme softness, cut from a Krupp gun, gave an ultimate tenacity of 72,000 lbs., which is very remarkable when the extreme softness of the specimen is borne in mind.

It might be inferred that the strongest quality of steel was always the best, but it is not so, the amount of tenacity which is desirable depends altogether upon the purpose for which it is to be used, the weaker and softer, or less steely qualities, being more tough, are preferred for many purposes, more especially for structures exposed to vibration.

The effect of tempering upon steel is to increase its strength, and, when the chill is rapid, to render it harder and more brittle. In tempering ordinary articles they are first heated to redness and then cooled in water, and thereby made over strong and over hard, and unfortunately very brittle; these defects are then subdued by the application of a gentle heat, which is continued or increased until the required degree of temper is attained.

The strength is determined by two things: first, the steeliness of the steel, that is, the proportion of carbon which it contains; and second, the rate of cooling. The highest degree of strength is obtained by selecting a high steel, heating it to a dull red, and then chilling it rapidly. These two conditions give a degree of strength combined with toughness, far beyond that which is obtained by giving a higher degree of heat, then chilling it suddenly, and afterwards reducing the temper.

When a block of mild cast steel is prepared for the interior barrel of a gun, it is desirable that all its properties should be known before it is put into the gun. Accordingly, two sets of specimens are cut from the block of steel, each set of specimens consisting of three pieces: the one set is used to find the best tempering heat for strength, elasticity, and ductility; the other set of specimens is used to find the best heat for toughness. One piece of each set is tried in its natural or untempered state; one of each is tried after being raised to a high temperature and dipped into oil, and one of each is tried after being raised to a less temperature and dipped into oil. The three specimens of the first set, prepared as above, are turned in a lathe for the testing machine, and their several properties carefully noted. Then the three pieces of the second set are taken to suitable apparatus, and bent backwards and forwards, in order to ascertain the toughness of each of the three pieces. When all is concluded and the results noted, a careful consideration is given to the whole; and after the advantages or disadvantages are summed up, the heat to be given to the gun-block is determined upon, and it is thereby brought to the highest conditions of strength, elasticity, ductility, and toughness, of which it is susceptible. Some specimens after tempering are found to stretch from 15 to 25 per cent. of their length, thus showing a degree of ductility which is surprising, and combining the qualities of steel and wrought iron in the gun-barrel.

The next Table shows the behaviour of gun steel in the testing machine.

Steel.

The following Table shows the mean results of a number of experiments carried out at Woolwich with cast steel of mild quality:—

TABLE X.

Nature of Material.	Weight applied in tons per square inch of section.	Diameter of specimen.	Length of specimen in inches.	Elongation in decimals of an inch. Visible.	Elongation in decimals of an inch. Permanent.	Elasticity as shown by the difference between the visible and permanent elongation.	Breaking weight per square inch of section in tons.
Soft cast Steel in natural state.	12·39 13·98 15·46 34·79	·533	2	·003 ·004 ·005 —	— ·001 ·003 ·372	·003 ·003 ·002 —	34·79
Cast Steel tempered at a high heat by immersion in oil.	15·01 19·78 20·46 25·01 25·92 32·06 32·29 33·66 53·78	·533	2	·0025 ·003 ·00325 ·0035 ·004 ·005 ·006 ·0085 —	— — — — — ·0005 ·0015 ·0035 —	·0025 ·003 ·00325 ·0035 ·004 ·0045 ·0045 ·005 —	53·78

As the result of a long course of valuable experiments made by Mr. G. Berkley, C. E., he draws the following general conclusions : 'First, that Bessemer steel will bear before rupture a minimum tensile strain of 33 tons per square inch of section, and stretch about 1 inch in 12 inches of its length; second, that the same material will bear, either in tension or in compression, a minimum stress of 17 tons, before the extensions or compressions per unit of stress become irregular or excessive, as compared with those which have preceded them, in other words, before the yielding point of the material is reached; third, that this material will probably contain about ·45 per cent. of carbon, chemically combined with the iron; and, fourth, that this description of steel, if properly made and annealed, is as uniform in quality as wrought iron, and may therefore be employed (precautions being taken to test its quality) as a substitute for wrought iron, while allowing an increase of strain of 50 per cent. to be imposed upon it.'

Different qualities of steel have varying ability to resist compression; with ten specimens of highly converted cast steel, of a quality suitable for cutting instruments, an average of 76,000 lbs. was required to compress the specimen $\frac{3}{1000}$th of an inch. With ten specimens of soft mild cast steel of the finest quality, the softest required 25,000 lbs., the hardest 46,000 lbs., or an average of 35,500 lbs. to produce the same impression; while two specimens, cut from a Krupp gun, required an average force of 25,300 lbs. One of these specimens was afterwards flattened down into a thin disc without cracking at the edges, thus showing a remarkable degree of malleability in the cold state. In all these experiments the specimens at the commencement were 1 inch in length and ·533 inch in diameter.

The material hitherto termed mild steel, and with which the experiments referred to in the previous pages of this chapter were made, was wholly produced from an almost

Steel.

pure wrought iron by the ordinary cementation process, much in the same indirect manner as the ordinary cast steel of commerce briefly described at page 68, but with the characteristic of containing a lower proportion of carbon.

The process employed in its manufacture, although it produced a fine material as regards quality, is necessarily expensive, and thus proves a drawback to its general use in the trades of manufacture and commerce.

During the past ten years so greatly an increased demand for such a mild steely material of a less expensive kind, affording the same good qualities together with uniformity, and so rendering it suitable for innumerable purposes in the various arts, has grown up, that some improved method of manufacture was most desirable.

Necessity, as ever, being the mother of invention, the late Sir W. C. Siemens devised an entirely new system, nearly as direct as that of the Bessemer, and producing as good a quality as that of the cementation, at a considerably lower cost than either.

Under such favourable conditions this new system of steel manufacture has been largely carried out, and the mild steel made thereby extensively applied to a variety of purposes for which the better qualities of wrought iron were formerly employed. At the same time, there have been marked improvements effected in the manufacture of Bessemer mild steel.

Certain of the cheaper sorts of crude cast iron contain a large percentage of phosphorus, which the Bessemer process failed to eliminate, and consequently could not be properly turned to account, as the presence of phosphorus greatly influences and deteriorates the quality of steel; now, however, the difficulty has been surmounted.

By a long course of experiments Mr. Gilchrist Thomas has solved the problem, and, by introducing a new basic lining, composed of magnesia and lime, into the Bessemer converter, has obtained this most satisfactory result—that the phosphorus is quite got rid of and the molten iron left comparatively pure.

Afterwards the spiegeleisen, containing a measured proportion of carbon, is added, and thus the iron is converted into steel.

It will thus be seen that at the present time there are three distinct systems adopted in the manufacture of mild steel, each possessing its own peculiar characteristics.

Firstly, the cementation process, namely, that of refining cast iron into an almost pure wrought iron, and then afterwards the combining of the pure wrought iron with a definite amount of carbon, during which the carbonic acid from the charcoal and oxygen is absorbed, and thus gives to the iron the nature of steel.

Secondly, the process introduced by Sir Henry Bessemer, in which a dense volume of air is forced through the crude cast iron in the molten state. During its passage the oxygen combines chemically with the carbon and passes away as carbonic acid, thus leaving the iron comparatively pure.

Thirdly, the Siemens process, which consists in mixing an iron ore rich in oxide with the mass of crude iron. In this, the oxygen of the ore performs the same chemical office as that from the air in the Bessemer process, and, uniting with the carbon in the crude iron, passes off as carbonic acid.

By means of the Siemens system manufacturers are enabled to produce steel of any degree of softness, suitable for steam boilers, rivets, ship-building and other purposes.

In order that the several properties of the Siemens steel should be known and recognised, a most carefully conducted series of experiments has been made, and the results duly reported to those societies immediately concerned in the matter, more especially to the Institution of Mechanical Engineers and the Iron and Steel Institute.

From analyses made by Mr. Edward Richards on three specimens of Siemens steel, the composition per cent. was found to be as under :—

Steel.

	No. 1.	No. 2.	No. 3.
Carbon	·192	·167	·234
Silicon	trace	·017	·013
Sulphur	·040	·044	·029
Phosphorus	·048	·050	·056
Manganese	·430	·457	·851
Copper	·021	·022	·024
Iron	99·269	99·243	98·793
	100	100	100

The relation of tensile strength per square inch to the proportional amount of carbon in these three specimens may be noted :—

No. 1 equal to 28·35 tons
No. 2 ,, 25·35 ,,
No. 3 ,, 32·24 ,,

One must, however, bear in mind that the ultimate strength of a particular specimen does not necessarily represent the exact cohesive force of the metal previous to experiment, because the specimen stretches and offers a smaller surface at the point of rupture than elsewhere.

The recent experiments with Siemens steel have been chiefly made with specimens of greater length than was the case with those formerly carried out at Woolwich (see page 13), and have thrown additional light upon the law of extension and elasticity of the metal under strain. The Siemens steel test pieces were 10 inches in length, and thus afforded a greater opportunity for measurement with microscopic accuracy. Had they been chosen 50 feet in length, as were those used in the experiments referred to at pages 57 to 63, the advantage would have been still more apparent.

Professor Kennedy observed that even with a difference of 2½ inches in length—7½ inches instead of 10 inches—the shorter were distinctly less uniform in extension than the longer specimens.

With the 10-inch specimens a decided set was perceptible, with a strain per square inch approaching to nine tons, or about 40 per cent. of the breaking load and about 60 per cent. of the load usually considered to be the limit of elasticity for practical purposes ; but the exact point at which set commences and elasticity ends is not easily found, and must accordingly be considered only as an approximation.

By being permanently strained up to fifteen tons per square inch, this steel does not seem to undergo any further change.

A 10-inch specimen was stretched ·01 inch by 17·34 tons, and afterwards by a continuous strain of fifteen tons, but without any perceptible alteration.

The following interesting experiment with a piece of soft Bessemer steel of excellent quality is instructive. The test pieces were first subjected to torsion and twisting, and then prepared for tensile strain, the one piece being annealed, the other unannealed. The results noticed were as under :—

	Annealed.	Unannealed.
Elastic limit	19·59 tons	29·88 tons
Tensile strain	31·01 ,,	37·77 ,,
Cohesive force	55·47 ,,	57·89 ,,
Elongation on 4 inches	26·25 per cent.	11·25 per cent.
Contraction of area	57·3 ,,	49·57 ,,
Mechanical work done for a length of 4 inches	28·94 inch tons	15·15 inch tons

The strength of both specimens had evidently been in-increased by the previous torsion ; but, of the two, the annealed would be the most reliable, although perhaps not so strong as the unannealed specimen.

Of late years much attention has been given to the chemical analyses of the crude iron out of which steel is now directly manufactured.

Both in the Bessemer and Siemens processes the different sorts of pig iron ore do not contain the same proportion of their several constituents, and for this reason a precise knowledge of the composition is necessary, in order that a proper uniformity in quality and strength may be obtained in the manufacture of the steel.

CHAPTER VII.

ON COPPER AND OTHER METALS, AND THEIR ALLOYS.

Copper.

THE reddish-brown sonorous metal termed copper is possessed of considerable strength and elasticity. It is a metal which is both malleable and ductile, but it possesses the former property in a higher degree than the latter; hence, it is much better fitted to be hammered, rolled, or worked into thin hemispherical pans, or sheets, or other forms, than to be elongated into a fine wire by pulling it through a drawplate. It is likewise to be noted that both properties, malleability and ductility, depend upon, and are mostly due to the purity of the metal.

Next to iron, in its various conditions, copper is one of the most important and useful metals found in the workshop, and is extensively employed in the mechanical arts. Its ores are widely distributed, and are found more or less abundantly in almost every country in the world. Copper probably has been longer in common use than any other metal.

As will be seen by the Tables in this chapter, its tenacity is somewhat uncertain, and is not at any time equal to that of either wrought iron or steel; but still as it is superior in strength to gold, platinum, or silver, and indeed to all the softer metals, and possesses many of their good properties besides, it is held in high estimation.

A square inch of good wrought copper will break with a tensile strain of about 15 tons ; but it is not so strong in the ingot or cast condition, for in that state it will often break with less than the half of the above tension. The effect of working upon copper is remarkable, both as regards its malleable and its ductile properties ; even a piece of good

G

copper wire, $\frac{1}{10}$th of an inch in diameter, may be worked up to such a condition, that it will require a strain of 300 lbs. to pull it asunder.

At page 120 of Mr. Bloxam's treatise on Metals, there will be found much useful information respecting the quality of copper, as depending upon the refining or toughening processes of its manufacture, by which it will be seen that it is altered, from a brittle state, into one which is soft, malleable, and ductile; but in the hands of the unskilful founder or furnace manipulator, the malleable copper, from mismanagement, may again become brittle by the absorption of charcoal, and, when such is the case, it has to be again refined by the action of air, while melted in the crucible of the founder.

It may be observed that, in the working of copper, either by drawing, rolling, or hammering, it is altered in the same way as wrought iron or steel; under the operation it becomes rigid, stiff, hard, and liable to crack, or even to disintegrate, and it can only be restored to its normal quality by a course of annealing, which may sometimes require to be several times repeated. This change is mechanical, and is quite distinct from the chemical change to brittleness before referred to.

Copper, even when employed by itself, is extensively used for manufacturing purposes, but the articles made of pure copper are few in number, when compared with the manifold forms in which it appears when combined with other metals. It is also the hardest as well as the most tenacious of all the workshop metals, except iron and steel, and it may be worked by the smith either cold or hot. When heated to redness, it can be forged, drawn down, or upset much in the same manner as wrought iron, but when it is heated to fusion or even to redness, and at the same time exposed to the atmosphere, it is found that the exterior surface is rapidly converted into black scales of peroxide. This may be roughly removed by heating and plunging into cold water, or the pure metallic surface may be laid bare by immersion in a solution of ammonia; but by repetition of the above process the copper may be entirely wasted.

In ordinary use it is very liable to corrosion, by the mere exposure of the bright parts of the metal to a damp atmosphere. The green oxide of copper so produced being poisonous, it is of importance that any copper vessels used for culinary purposes should be kept perfectly clean; the usual precaution is, to cover the interior surface of the copper with a coating of tin, but this does not afford perfect security, as, from its softness, it is liable to be scraped or otherwise worn off, by the tear and wear of daily use.

Copper has been used to a considerable extent for the construction of steam boilers, more especially for marine purposes, or where salt water has to be evaporated. But, of late years, it has been entirely superseded by iron or steel. Although copper does not ordinarily rust to the same extent by the action of salt water, still it is more rapidly damaged in the furnaces by the use of sulphurous coal, and when the boiler does happen to leak from any cause, the leakage does not take up so readily as with iron, and the salt incrustation is found to deteriorate the metal more, in the vicinity of the leakage; besides, copper is sooner reduced in tenacity by over-heating. Iron does not become perceptibly weaker up to a temperature of 570°, after that temperature is reached, the strength gradually decreases; whereas copper is in its best state when cold, and loses tenacity by every increase of temperature. With so many disadvantages, and having a tenacity of only 15 tons, it is now disused, and iron or steel is now almost invariably employed for steam-boilers, except in some special cases.

The specific gravity of copper has a considerable range; it varies from 8·78 in the crude state to 9·0, after rolling or hammering. An increase of density was supposed to arise from the mere condensation of the particles of the mass, but it is now suggested that when copper is melted in contact with the atmosphere, it absorbs oxygen, which does not afterwards find the means of escape, and consequently the metal becomes slightly porous; this absorption of oxygen is in a measure prevented by careful fusion under common salt. The density of the metal, so fused, is nearly equal to that of worked copper, namely 8·921, and after being subjected to a pressure of 300,000 lbs. per square inch, it increases to 8·930; this difference however is small, and it has been suggested that the change may be more owing to a diminution of the empty spaces still remaining, rather than to the approximation of the molecules.

Articles which, from their form, require to be cast in a mould, are seldom made of pure copper, because it is too soft for the generality of purposes; it is likewise very porous after casting, and has great tendency to unsoundness. Copper melts at 2,000°, which is higher than the melting-point of most of its alloys, as given in Table XVII.

The ultimate tenacity of copper when in the cast state ranges from 19,000 to 26,000 lbs. per square inch of section; but the strength of this copper may be considerably increased by working, wire-drawing, hammering, or consolidating. When carefully drawn into wire, copper has a tenacity as high as 60,000 lbs. per square inch; ordinary copper bolts a tenacity of 33,000 lbs.; while several other specimens, of pure wrought copper bolts, gave an average ultimate tenacity of 36,000 lbs. per square inch.

Copper, in castings, is generally mixed with other metals; such mixtures of metals are called alloys, and those alloys in which copper predominates are the most numerous; tin, zinc, lead, phosphorus, aluminium, iron, and other metals are all employed to form alloys with copper.

The fluidity and tenacity of copper may both be considerably increased by the addition of a small percentage of phosphorus. In several experiments that were made, this alloy was remarkable for its density and tenacity, and when broken showed a regular, sound, and uniform fracture Mr. Abel found that by an addition of from 2 to 4 per cent. of phosphorus, a metal was obtained more uniform than bronze, and having an ultimate tenacity of from 48,000 to 50,000 lbs. per square inch. Mr. Overman states that by the addition of phosphorus, copper may be rendered as hard as steel. The combination of phosphorus with copper increases the tendency to corrosion, unless another metal, such as tin, is also added. By adding phosphorus to bronze, its homogeneity is materially increased, and its tendency to oxidise by exposure to the atmosphere completely neutralised.

The two following Tables show the results of experiments with exceptionally good copper, and with copper containing a small percentage of phosphorus.

Results of experiments made at Woolwich, to ascertain the tensile strength of copper :—

TABLE XI.

Length of specimen under test, 2 inches.

Breaking weight in tons per square inch of section.	Specific gravity.	Remarks.
15·39 14·78 14·89 11·79	8·688	Cast copper (very pure).
15·75 16·25 15·9 16·25		Specimens cut from virgin copper bolts. The bolts were 2½, 2, 1, and ¾ inch diameter respectively, and the diameter of the specimens cut from the bolts, 1″, 1″, ·6″, ·6″.

The above results, especially those on cast copper, indicate a better quality of copper than the average.

Results of experiments made at Woolwich to ascertain the tensile strength of phosphorised copper :—

TABLE XII.

Length of specimen under test, 2 inches.

Breaking weight in tons per square inch of section.	Specific gravity.	Percentage of Phosphorus.	Diameter of specimen in inches.
6·23	—	1	1·13
7·56	8·202	1	1·13
16·47	8·592	1	1·13
17·13	8·876	1½	1·13
19·0	—	1	1·13
20·25	8·614	2	1·13
20·34	—	1	1·13
20·41	8·580	2	·98
21·27	8·615	2	1·13
21·38	8·422	3	1·12
21·5	—	—	—

The above nine specimens show an average tenacity of about 20 tons per square inch. The first two specimens are not included.

Bronze and Gun-metal.

The alloy of copper and tin, usually termed bronze, but sometimes gun-metal, has been of great value and importance in the arts, from time immemorial. When this alloy is used for artillery purposes, it generally consists of from 90 to 90½ parts of copper, and from 9½ to 10 parts of tin. A small fraction of the tin, however, is invariably lost in the melting process, even if performed in crucibles, still more is lost when a reverberatory furnace is employed, and most of all if the fusion is made in a cupola. When copper and tin are alloyed in proper proportions, a harder metal than either is produced, with a density greater than the mean density of the constituents. The alloy is more fusible than copper, and less liable to corrosion.

Copper and tin mix well in almost all proportions, a

small percentage of the latter rendering the alloy both hard and tenacious, and by changing the proportions of tin to copper, alloys are formed, varying extremely in colour, hardness and soundness, and also in regard to their capability of yielding a sonorous sound when struck, as when used for bells.

By the addition of tin, hardness is increased. With a proportion of ⅛th of tin to ⅞ths of copper, by weight, the metal assumes its maximum hardness for the purposes of the engineer, or for any application in which it requires turning or planing. The founder sometimes resorts to a mixture containing from ¼th to ⅓rd of tin, it then becomes highly brittle and elastic, like glass, and the sonorous property is improved in a high degree; at the same time its brittleness rather than its hardness is the prominent feature. A mixture, consisting of 2 parts of copper and 1 of tin, forms an alloy so hard that it cannot be cut with steel tools, and has a highly crystalline structure. At this stage almost every characteristic for which tin and copper are distinguished seems to be entirely changed.

The specific gravity of ordinary bronze varies from 8·4 to 8·94, according to the nature of the mixture and the way in which it has been treated by the founder. From Muschenbroek's experiments, it appears, that the density of the alloy becomes greater, as the proportion of tin is increased: with 10 parts of copper and 1 of tin, the specific gravity was 8·351; with 8 parts of copper and 1 of tin, it was 8·392; with 6 parts of copper and 1 of tin, it was 8·707, and with 4 parts of copper and 1 of tin, it was 8·723. The results obtained in experiments made at Woolwich showed a rather lower specific gravity than the foregoing, and Major Wade states that, judging from an examination of the specimens obtained from the heads of all the guns cast at an American foundry, the density varied from 8·308, to 8·756, thus showing a difference of weight between the lightest and heaviest specimens of 28 lbs. in the cubic foot.

Bronze, melting at 1,900° Fahr., is more fusible than copper, but much less fusible than tin. Indeed, the difference of fusibility in these two metals is so great, as to render it exceedingly difficult for the founder to obtain a perfectly homogeneous bronze alloy, when the casting is in great mass; upon examination it is often found that specimens of large castings show a great want of regularity, and contain patches or spots of the appearance of tin, mechanically interposed between the particles of alloy. These spots, although apparently of tin, seldom contain more than 25 per cent. of that metal.

In forming bronze, great care must be exercised by the founder, when mixing the metal in the furnace. The more refractory metal should be melted first, and the more fusible metal added. The alloy should be well stirred and then cast, and should be cooled as rapidly as possible, in order to obtain uniformity and compactness, and to obviate the tendency to separation in the process of cooling, the denser metal being generally found at the lower part of the casting, and the lighter one at the top, if the cooling is long protracted.

When mixing tin with copper in the furnace, it is of the utmost importance that the tin should not be long exposed to the influence of the air, because, when it is heated to redness, its affinity for oxygen is so great that it will be rapidly oxidised and converted into the peroxide or putty of tin (the putty-powder of commerce). This is very disadvantageous, for when a reverberatory furnace works slowly, the metal is found to contain innumerable particles of the putty-powder, and to such an extent is this sometimes the case, as to render the turning or boring of the gun a work of extreme difficulty, because the cutting instruments are blunted by the hard oxide. Hence it is that the crucible system of melting has a great advantage over the large reverberatory furnace, wherever it is applicable, the chief barrier to the use of the crucible being the difference in cost.

Copper and other Metals, and their Alloys. 89

In mixing copper and tin to form gun-metal, the best arrangement is to alloy them first in the proportion of 2 to 1, obtaining a white hard crystalline silvery speculum metal, and then by another melting to mix this hard alloy with the requisite quantity of copper to give the required quality of bronze. In this second melting, as well as in the first, the copper is first melted; then, shortly before the time of casting, the 2 to 1 alloy is added, and the whole well stirred to secure a good mixing. It is afterwards immediately cast in order to prevent the oxidation of the tin.

A small quantity of zinc is sometimes added to common bronze, for the purpose of making it mix better. Zinc increases the malleability without materially reducing the hardness, but it is seldom used in gun-metal.

The investigation bestowed upon the strength and other properties of bronze, during the last hundred years, has been most thorough; many experiments have been made, but the results are so discordant as to render it difficult to give an account of them without inserting such a number of tables as would be incompatible with the size of this volume.

The late Mr. Rennie came to the conclusion, that the average ultimate tenacity of gun-metal is about 36,333 lbs. per square inch, that is, when the mixture consists of copper and tin only, and in the usual proportions. Muschenbroeck's experiments go to show that with a mixture of 6 parts of copper and 1 part of tin, the ultimate tenacity is equal to 44,000 lbs. per square inch. From similar experiments made at Woolwich, the following results were obtained, with a mixture of—

Tenacity.
12 parts of copper and 1 of tin, 29,000 lbs. per sq. inch.
11 ,, ,, 30,700 ,, ,,
10 ,, ,, 33,000 ,, ,,
9 ,, ,, 38,000 ,, ,,

The above results are not quite uniform, but they nearly agree with those obtained both by Rennie and Muschenbroeck, and are perhaps not far from the average of the bronze usually met with in the arts.

Judging from an immense number of experiments made at Woolwich, at different times, during the past fifteen years, and without having regard to many minor points, or to the shades of proportion of mixtures and different modes of casting, the average tenacity of bronze is 31,280 lbs., varying from 22,500 lbs. to 41,000 lbs. per square inch. As a rule, it begins to yield visibly, and to take permanent set at 15,164 lbs., and the average ultimate elongation per inch in length is ·290 of an inch.

Some experiments are recorded, both much lower and much higher than the above, even ranging from 17,698 lbs. upwards to 56,786 lbs., but these are exceptional, and 33,000 lbs., which is between 14 and 15 tons, may be considered as the general average of good bronze.

Major Wade, of the United States Army, has given great attention to this point, and he has found the ultimate tenacity to vary from 23,929 lbs. to 35,484 lbs. per square inch, a difference in the ratio of 2 to 3. These differences occurred in samples taken from the same part of different castings of gun-heads, where the materials used were apparently all of the same quality, and were melted, cast and cooled in the same manner, and every effort was used to have them similarly treated in all respects; but the causes of such irregular and unequal results, when the materials used and the treatment of them were apparently the same, are very obscure and perplexing.

Benton states that the quality of bronze depends much upon the nature of the furnace treatment of the melted metal, and that by extreme care, and by this alone, the tenacity of bronze made at Washington Navy Yard Foundry has been raised as high as 60,000 lbs. per square inch, so as to equal in tenacity good wrought iron, which is worth noting.

The following results have been obtained by analysing a number of tables which show the tensile strength of various specimens taken from different parts of bronze guns, some of the specimens being cast in open moulds with both green sand and dry sand; others in closed moulds with green sand, and the remainder in iron moulds, forming large chills to cool the metal rapidly.

The average weight in lbs. per square inch, required to overcome the elasticity of the metal was 14,694 lbs., and the point of fracture 27,305 lbs. per square inch, the average elongation per inch in length being ·144 inch. Each specimen was 1·066 inch in diameter, and the length of breaking part 2 inches.

In sixteen specimens, each ·533 of an inch in diameter by 1 inch in length, the compressive stress per square inch required to produce a permanent set of ·001 ranged from 5·4 tons to 7 tons, and was usually near to a mean between these two extremes. A stress of 7 tons applied to each specimen produced a permanent set ranging from ·0015 to ·0005, the most elastic having been cast in an iron mould which cooled the metal rapidly. Taken as a whole, the bronze when in compression, showed a permanent set of ·001 of an inch with an average compressive load of 13,843 lbs. per square inch, and ·384 with 50 tons per square inch, with which load the specimens were fractured. When compared with steel, or even iron, bronze offers a small resistance to compression; but its compressibility is found to depend greatly upon the perfection of the alloy, and the amount of fluid pressure induced by the height of the deadhead, and the rate of cooling employed to prevent the separation of the tin.

Brass.

Of the many useful alloys known to the mechanical world, brass is perhaps the one which is most extensively used, being easily worked, and of a fine yellow colour. In this alloy as in nearly all the others used in construction, copper is the most prominent metal.

In the melting and mixing of copper and zinc, great care has to be exercised to prevent the zinc from passing away in vapour, which is usually effected by covering the crucibles with charcoal powder and a close lid of clay. Brass is more malleable than copper when in the cold state, but it will not submit to be forged at a red heat, on account of the low melting point of zinc; even a small addition of zinc to copper will affect it in this important respect, which is a disadvantage.

Tin and lead are sometimes added to copper and zinc in making brass. The quantity of copper varies from 60 to 92 per cent., and it is not uncommon to add from $\frac{1}{2}$ to 3 per cent. of lead, and from $\frac{1}{4}$ to 3 per cent. of tin, according to the nature of the work for which the alloy is required. The best proportion for fine or yellow brass appears to be copper 2 parts and zinc 1 part.

The specific gravity of brass ranges from 7·82 to 8·5, and is thus greater than the mean of its constituents, that of copper being 8·78 and of zinc 6·86, which is probably due to the zinc finding empty spaces, into which it enters.

The melting point of brass is lower than the mean of its constituents would indicate, being from 1689° Fahr. to 1900° Fahr., the fusibility entirely depending upon the quantity of zinc which has entered into combination with the copper; and the fact of the fusibility of the alloy increasing with the further addition of zinc furnishes strong evidence that the change in their properties, which the metals undergo, arises from chemical affinity. The liquidity of this alloy may be very much increased by adding a small portion, say half an ounce, of dry phosphorus in the crucible, and then stirring the metal before running it into the mould; by this means the alloy is rendered so liquid that very thin and sound castings may be obtained without difficulty.

The addition of a little lead causes brass to be more ductile and better adapted for turning in a lathe than common brass, while a large addition of lead renders it very brittle, and if the proportion of lead amounts to nearly one half, then a partial separation takes place in the act of cooling. The tenacity of fine brass (as stated by Rennie) is only about 18,000 lbs. per square inch; but the average ultimate tenacity of the best quality of brass, when composed of two parts copper and one part zinc, is much higher, being 28,900 lbs. per square inch.

Muntz-metal.

The alloy of 60 parts copper and 40 parts zinc, termed 'Muntz-metal,' is a mixture which has been much used for the sheathing of ships, and for the bolts of marine engines liable to rust, and for similar purposes. Its tenacity is about 22 tons per square inch, or nearly equal to that of good wrought iron, while its endurance in salt water is nearly equal to that of bronze or brass.

In consequence of the high tenacity of this metal, it was thought desirable to cast guns of it, but it was found that, during the increase of temperature due to rapid firing, the presence of zinc affected the tenacity of the metal to a considerably greater extent than is the case with ordinary gun-metal, and, notwithstanding its great strength, it was found unsuitable.

The following Table shows the results of various experiments made at Woolwich to ascertain the tensile strength of brass, &c. :—

TABLE XIII.

Length of specimen under test, 2 inches.

Breaking weight in tons per square inch of section.	Specific gravity.	Diameter of specimen in inches.	Mixture.
16·02	8·58		Copper, 7 lbs., zinc, 3 lbs. 8 ozs.
16·02	8·534		
13·35	8·699	1·13	
10·57	—		—
10·68	—		
10·91	—		
3·12	7·049	1·3	Copper 10 parts, iron 10 parts, zinc 80 parts.
3·22	6·915		

The average ultimate tenacity of the first 6 specimens was 12·92 tons or 28,940 lbs. per square inch. The result of the last two experiments is instructive.

Sterro-metal.

'Sterro-metal' is a new alloy of copper, zinc, tin, and wrought or pure iron. One of the principal objects for which this particular kind of metal was first introduced was, to supersede the use of cast and wrought iron, and bronze, in the manufacture of heavy ordnance. Cast iron is objectionable on account of its low tenacity, elasticity, and ductility, and has proved not altogether suitable for the purpose. Wrought iron is rather too soft for the interior of guns; necessitates, in its manufacture into guns, a great amount of expensive working, and, even when the greatest care and skill is exercised, its perfect soundness cannot be altogether depended upon. Bronze, when of the proper quality, is too soft; its ultimate tenacity is uncertain, being generally not much greater than that of good cast iron; it is more easily stretched to the elastic limit than is the case with good wrought iron; and, in addition to all these objections, it is more costly.

For the foregoing, as well as for other reasons, many efforts have been made to obtain a homogeneous and dense metal, which could be cast into large masses, and which combines all the good properties of hardness, great tenacity, elasticity, and soundness.

Sterro, or firm metal, was first introduced to the arts in Vienna, a few years ago, by Baron de Rosthorn, and was rapidly seized hold of by experimentalists in various parts of the world. At Woolwich a series of interesting experiments were made, with specimens containing different proportions of the various metals, prepared in the Royal gun factories, as well as with specimens of the metal obtained from Austria. Since that time additional experiments have been made, and the results are given in the following Tables.

From these it will be seen that this alloy has some most valuable properties, especially stiffness, as its name implies, together with great tenacity and power of resisting compression, with considerable hardness, which is very desirable in any gun-metal, to resist the abrasive effect of the projectile. Although the sterro-metal possesses the most prominent of these qualities in a high degree, yet, even with this metal, it is very difficult to ensure a perfectly uniform and sound casting with any degree of certainty, several fractured specimens showing the mixture of metals to have been very incomplete.

From the following Tables it will be seen that the ultimate tenacity varies from 43,000 lbs. to 85,000 lbs. per square inch, or an average of 60,480 lbs. per square inch; it also required a strain of 30,000 lbs. to produce a permanent elongation of ·002 of an inch per inch of length, while the ultimate elongation with the average strain of about 60,480 lbs. per square inch was ·0675 of an inch per inch of length. Whereas, bronze begins to yield at 15,000 lbs., and is fractured at 33,000 lbs. per square inch, and shows an ultimate elongation of ·290 of an inch per inch of length.

The recent experiments made with steel, and especially with steel tempered in oil, and the practical success which has attended it for gun purposes, has probably barred the way to further experiments being made with sterro-metal, at least for the present; but it will be evident that the subject opens up a wide field for other experiments with mixtures of iron and other metals, with a view to discover other combinations which will give strength with other good qualities.

Results of experiments made at Woolwich to ascertain the tensile strength of sterro-metal:—

TABLE XIV.

Diameter of specimen, ·707 inch; length of part of specimen under test, 2 inches.

Breaking weight in tons per square inch of section	Strain at permanent elongation of ·002 in. per inch of length in tons.	Ultimate elongation at breaking point in inches.	Treatment.	Mixture.
26·75	6·75	·1	as received.	Austrian.
21·5	11·0	·05	} cast in sand.	Copper 60, zinc 39, iron 3, tin 1·5.
19·25	13·75	·015	cast in iron.	
24·25	17·25	·016	{ cast in iron and annealed.	Copper 60, zinc 44, iron 4, tin 2.
23·25	15·25	·02		
28·0	17·0	·045	forged red hot.	
31·61	—	—	{ cast in iron and forged red hot.	— —
32·52	—	—		
34·0	—	—	—	Copper 60, zinc 37, iron 2, tin 1.
38·0	—	—	—	Copper 60, zinc 35, iron 3, tin 2.
27·0	—	—	after simple fusion.	Copper 55·04, spelter 42·36, iron 1·77, tin ·83.
34·0	—	—	forged red hot.	
38·0	—	—	drawn cold.	
28·0	—	—	after simple fusion.	
32·0	—	—	forged red hot, drawn cold and reduced from 100 to 77 transverse sectional area.	Copper 57·63, spelter 40·22, iron 1·86, tin 0·15.
37·0	—	—		

The next Table shows the results of experiments made at Woolwich to ascertain the extension and tenacity of sterrometal:—

TABLE XV.

Diameter of specimen, ·707 inch; length of part under test, 2 inches.

1 Breaking weight in tons per square inch of section.	2 Weights applied in tons per square inch of section.	3 Elongations in decimals of an inch due to the weights applied in column 2.	4 Permanent elongations in decimals of an inch due to the weights applied in column 2.	5 Permanent elongations per inch of length in decimals of an inch.	6 Treatment.
20·35	2·27 3·86 6·02 7·27 8·98 10·40 11·93 12·50 13·07	·001 ·002 ·003 ·004 ·005 ·006 ·007 ·008 ·009 ·074	·001 ·0015 ·002 ·0035 ·004	·0005 ·00075 ·001 ·00175 ·002	Cast in sand.
19·78	4·09 5·68 6·48 6·93 7·73 8·30 9·21	·001 ·002 ·003 ·004 ·005 ·006 ·008 ·227	·0005 ·001 ·0025 ·003 ·005	·00025 ·0005 ·00125 ·0015 ·0025	
20·92	5·11 7·16 8·07 8·41 8·75 9·32	·002 ·004 ·005 ·006 ·007 ·009 ·282	·001 ·0025 ·003 ·004 ·0055	·005 ·00125 ·0015 ·002 ·00275	
19·21	4·54 6·36 10·12 12·50	·002 ·004 ·006 ·009	·0005 ·001 ·0025 ·0045	·0025 ·0005 ·00125 ·00225	
20·35	7·50 11·37 13·19 14·66	·003 ·005 ·007 ·009	·0005 ·001 ·002 ·004	·0025 ·0005 ·001 ·002	

TABLE XV.—*continued.*

1 Breaking weight in tons per square inch of section.	2 Weights applied in tons per square inch of section.	3 Elongations in decimals of an inch due to the weights applied in column 2.	4 Permanent elongations in decimals of an inch due to the weights applied in column 2.	5 Permanent elongations per inch of length in decimals of an inch.	6 Treatment.
32·52	5·23 8·64 14·10 15·46 17·05 18·53	·002 ·004 ·006 ·007 ·008 ·01	 ·0005 ·001 ·0025 ·003 ·167	 ·00025 ·0005 ·00125 ·0015 ·0835	Cast in sand and then forged.
30·02	5·57 9·66 13·75 16·26 16·94	·002 ·004 ·006 ·008 ·01	 ·001 ·0025 ·0035 ·167	 ·0005 ·00125 ·00175 ·0835	
20·35	4·32 6·25 8·07 9·21 10·23 10·91 11·14 11·82 12·28	·001 ·002 ·003 ·004 ·0055 ·0065 ·007 ·008 ·01 ·132	 ·0015 ·0025 ·003 ·0045 ·005	 ·00075 ·00125 ·0015 ·00225 ·0025	Cast in iron (specimens defective).
21·26	5·23 8·52 10·57 11·14 12·05 12·50	·001 ·003 ·005 ·006 ·008 ·01 ·156	 ·001 ·002 ·0045 ·005	 ·0005 ·001 ·00225 ·0025	

TABLE XV.—continued.

1	2	3	4	5	6
Breaking weight in tons per square inch of section.	Weights applied in tons per square inch of section.	Elongations in decimals of an inch due to the weights applied in column 1.	Permanent elongations in decimals of an inch due to the weights applied in column 2.	Permanent elongations per inch of length in decimals of an inch.	Treatment.
27·06	4·20 6·02 8·52 10·34 12·73 14·32 14·78 17·28 18·76	·001 ·002 ·003 ·004 ·005 ·006 ·007 ·009 ·012	·0005 ·001 ·0025 ·0045	·00025 ·0005 ·00125 ·00225	Cast in iron.
24·56	6·36 9·21 10·57 11·82	·002 ·004 ·006 ·009	·0005 ·0025 ·0045 ·275	·00025 ·00125 ·00225 ·1375	
25·69	7·84 9·55 13·19 14·78 15·46 17·39 18·53	·003 ·004 ·005 ·006 ·007 ·008 ·0105	·0005 ·0005 ·001 ·0025 ·0035 ·037	·00025 ·00025 ·0005 ·00125 ·00175 ·0185	
32·29	3·52 6·82 10·91 12·28 15·92 16·37 17·28 19·33 21·83	·002 ·004 ·006 ·007 ·008 ·009 ·01 ·012 ·0155	·0005 ·0007 ·001 ·0025 ·005 ·12	·00025 ·00035 ·0005 ·00125 ·0025 ·06	Cast in iron and then forged.

On the Strength of Metals.

TABLE XV.—*continued.*

1	2	3	4	5	6
Breaking weight in tons per square inch of section.	Weights applied in tons per square inch of section.	Elongations in decimals of an inch due to the weights applied in column 2.	Permanent elongations in decimals of an inch due to the weights applied in column 2.	Permanent elongations per inch of length in decimals of an inch.	Treatment.
27·51	3·18 5·68 8·87 9·78 12·05 12·50 13·30 15·01 17·05	·002 ·004 ·006 ·007 ·008 ·009 ·01 ·012 ·0155	·0005 ·0007 ·001 ·0025 ·0045 ·11	·00025 ·00035 ·0005 ·00125 ·00225 ·055	Cast in iron and then forged.

Results of experiments made at Woolwich to ascertain the transverse strength of mixtures of wrought iron and gunmetal:—

TABLE XVI.

Size of Specimen.			Distance between supports in inches.	Deflection in inches.	Breaking weight in tons at centre of beam.	Mixture.
Length in feet.	Breadth in inches.	Depth in inches.				
2	1·99	2	20	·142	2·8	equal proportions.
2	2·03	2	20	·115	2·74	
2	1·99	2	20	·110	2·24	
2	2·03	2·05	20	·110	2·56	gun-metal 1, iron 3.
2	2·02	2·02	20	·105	2·16	
2	2·02	2·05	20	--	1·31	
2	2·04	2·03	20	·140	3·0	gun-metal 1, iron 7.
2	2·01	2·06	20	·133	2·83	
2	2·01	2·05	20	·120	2·11	

It is right to state that the fracture showed the mixture of the two metals to be very incomplete; nevertheless, the result is worth noting, as it is probably the only Table of the kind.

Aluminium Bronze.

Aluminium Bronze is an alloy of copper with aluminium, which seems to promise great results. It consists usually of 90 parts of copper to 10 of aluminium, these proportions give an alloy which may be forged either cold or hot, in the same manner as wrought iron, but which does not weld. It is highly malleable and ductile, possessed of great stiffness and elasticity, its specific gravity being nearly the same as that of wrought iron; and it does not readily tarnish by exposure to the atmosphere.

From a number of experiments made at Woolwich, its average tenacity was found to be 73,185 lbs. per square inch, and its maximum tenacity 96,320 lbs., thus its strength was more than double that of bronze, greater than that of wrought iron, or even of many of the mild qualities of cast steel, and at the same time its resistance to compression exceeded that of ordinary cast iron. Colonel Strange, who has given much attention to this alloy, states that its rigidity is three times that of bronze, and many times greater than that of brass, and that it is less affected by changes of temperature than either of those alloys. In the liquid state, it can be cast into any form without difficulty, it works nicely under the file or in a lathe, and its other advantages are numerous.

In making this alloy, extremely pure copper has to be used, as with impure copper the alloy is much deteriorated in all its good properties; and to produce the best results, it requires to be remelted several times; the fact of remelting being required, in order to develope its greatest strength and stiffness, is not peculiar to this alloy, but is the case with cast iron and some of the other alloys.

The chief barrier to the extensive use of aluminium bronze is the cost of the aluminium, which makes the price of the bronze at least four times that of ordinary gun-metal.

Hence this metal is chiefly used for purposes where strength and stiffness, combined with lightness and non-liability to rust, aret he chief objects, as for surveying instruments, which have to be light and strong, and not liable to injury when carried from place to place, and used in tropical climates.

Other Alloys.

Bells are usually made of copper and tin of various mixtures, but generally in about the same proportions as for gun-metal. The smaller class of bells have a greater proportion of tin, to give hardness, and sometimes a little zinc; even silver is occasionally added, in order to improve the tone.

For such purposes, the question of strength is unimportant. The same remark applies to the usual bronze for statues, which is a similar mixture to gun-metal, but analysis of different statues shows considerable variation in the proportions of the two metals. The addition of a small portion of phosphorus would have the effect of preserving the surface of the metal from the influence of the atmosphere, which fact should be noted.

Babbitt's metal for machinery bearings, which consists of about 4 parts of copper and 8 lbs. of regulus of antimony and 96 lbs. of tin, is not a strong alloy, but it is one of the best reducers of friction, and its want of stiffness has to be provided for in the iron casing in which it is contained when in use, to form a bearing for a moving spindle. The casing is required in order to prevent it from spreading outwards under the pressure to which it is exposed. In the use of Babbitt's metal for bearings, it is necessary to exercise great care to prevent the heating of the journal. If this takes place, the Babbitt's metal melts and runs out.

Table showing the average melting point of most of the ietals and alloys referred to in this chapter.

TABLE XVII.

Metals and Alloys.	Melting point in degrees Fahrenheit.
Cast iron, rich in carbon	2786
,, 2½ to 3 per cent. of carbon	2600
Wrought iron	3280
,, freed from silicon	3500
Cast steel, mild, ·28 per cent. of carbon.	3300
,, ordinary, ·56 per cent. of carbon	3000
,, high, ·75 to ·8 per cent. of carbon	2850
Copper	2000
Tin	442
Zinc	758
Gun-metal	1900
Brass	1847
Lead	600
Antimony	800
Bismuth	483
Silver	1873
Platinum	3280
Gold	2016

ADDITIONAL EXPERIMENTS.

The following Tables show the result of certain experiments on the resistance of metals.

Resistance of Metals to Tension.

Size of specimen.	Nature of material.	Stress in tons per square inch at Yielding.	Stress in tons per square inch at Breaking.	Elongation per inch.	Elongation per cent.	Authority.
Diameter ·533″ area ·223 sq. in. Length of breaking part 2 inches	Homogeneous steel, soft,	12·8	31·9	·123	12·3	Woolwich Testing Machine.
	Do. tempered in oil at a low heat,	29·7	47·9	·129	12·9	
	Do. tempered in oil at a high heat,	29·9	48·5	·08	8	
Diameter ·754″ area ·4465 sq. inch Length of breaking part 2 inches	Best gun-iron, soft,	11·5	22·2	·134	13·4	
	Bronze	6·9	15.7	·162	16·2	
	Copper	—	15·5	—	—	

Resistance of Metals to Compression.

Size of specimen.	Nature of material.	Yielding strain in tons per square inch.	Total compression with load of 50 tons allowed to remain on for 5 minutes per inch.	Total compression with load of 50 tons allowed to remain on for 5 minutes per cent.	Authority.
Diameter ·533″ Length 1″	Soft steel	12·47	·218	21·8	Woolwich Testing Machine.
	Tempered steel	28·98	·054	5·4	
	Wrought iron, soft	12·2	·274	27·4	
	Bronze	7·5	·3423	34·23	
	Cast iron	15·9	·0308	3·08	

One marked feature in the bronze of modern times is a combination of copper with iron and manganese to form a most useful alloy called manganese-bronze, which has been brought to its present condition by Mr. Parsons, of London, and fulfils most important functions.

In the preparation of this bronze a suitable proportion of ferro-manganese is added to molten copper, the former being melted in a separate crucible, and the mixture thoroughly stirred in order to secure uniformity.

Due to its affinity for oxygen, the manganese cleanses the copper of oxides, and with them rises to the surface as slag.

By a variation in quantity of the component metals, several different alloys, each possessing some useful characteristic, such as strength, toughness, hardness, and ductility, are obtained; while some of the mixtures rival mild steel with regard to elasticity and toughness.

On account of its strength and non-corrosive nature, the manganese-bronze is well adapted for the manufacture of screw propellers for steamships.

It may be cast in sand, admits of being rolled into plates or other forms, or can be compressed in the molten state, just as steel on the system introduced by Sir Joseph Whitworth, referred to at page 71, and with similarly satisfactory results.

The following tables will suffice to exemplify some of the properties of the bronze :—

TESTS OF MANGANESE-BRONZE.

By Tensile Strain.

Description.	Elastic limit, tons per sq. in.	Breaking strain, tons per sq. in.	Ultimate elongation per cent.	Remarks.
No. 1. Rods, rolled hot	11·00 13·17 23·54 24·32 34·40	29·00 29·29 31·60 31·43 39·60	44·6 33·4 26·5 23·3 11·6	Mild, annealed for riveting cold Annealed. } As delivered from the rolls. Ditto, and finished cold.
No. 1. Plates, rolled hot	14·06 14·06 14·80 16·70	28·46 30·13 30·78 30·10	23·2 47·8 34·1 28·8	Pulled across fibre ,, with fibre ,, across fibre } Annealed ,, with fibre
No. 2. Cast under pressure	18·00 16·23	35·00 31·90	22·0 12·4	Cast in an iron cylinder and pressed when liquid.

By Torsion.

Description.	Diameter, ins.	Twisting moments in inch pounds. Elastic limit.	Twisting moments in inch pounds. Breaking strain.	Amount of twist in length of one diameter. No. of turns.	Remarks.
No. 2, cast under press.	·622	1,170	3,360	·183	Uniform twist.
	·624	1,200	3,372	·166	,,
*No. 1, rolled	·621	1,110	2,880	·175	Annealed.
†Rod . .	·621	1,980	3,242	·165	Rolled hot and tested as it came from the rolls.

* Was removed from machine unbroken.
† Was broken, showing a clear shear.

CHAPTER VIII.

TIMBER.

ALTHOUGH it is of less importance to investigate the strength of timber at the present time than it was formerly, in consequence of the diminished use of that material in permanent structures, and the more general employment of iron, still it will always be a very valuable material for certain purposes, and ought not to be neglected. Timber is variously used, even now, in permanent works, and is applied much more extensively in temporary structures—such as centerings and scaffolding. Hence its properties are well worthy of careful attention; and the student should be familiar not only with the external appearance of the principal kinds of wood, but also with their relative strength, stiffness, toughness, and durability.

One of the most obvious inferences to be drawn from the experiments recorded in the previous chapters is that very wide variations exist in the strength and other elastic properties of different metals, and even of different specimens of the same metal. If we could investigate the properties of timber with the same care which has been bestowed on the metals, we should find that there is an even greater variation in the properties of different kinds of wood. This arises in part from the fact that timber is much affected by a number of external and internal conditions, during its growth and seasoning, and in its subsequent treatment, which gradually modify and change its properties.

It will only be necessary in this chapter, to treat of the powers of resistance of a few among the many kinds of wood now employed in the mechanical arts. The greater number of the varieties of wood owe their commercial value to special characteristics, such as beauty of grain and capability of being polished—the description of which does not fall within the scope of the present treatise.

As a general rule, we may judge of the hardness of a wood by its specific gravity, if it is in its natural state. But the density may be increased by artificial compression (as in the manufacture of trenails) and this increase of density is generally accompanied by increase of strength. Some varieties of wood, as, for instance, *lignum vitæ*, are so dense that they sink in water, while some of the softer woods have not half the density of that fluid. The presence of gum or resin in any wood adds both to its strength and durability. Many woods will last a long time, if kept constantly under water, but scarcely any wood is very durable when allowed to become wet and dry alternately.

The strength of a piece of timber depends upon the part of the tree from which it is taken. Up to a certain age, the heart of the tree is the best ; after that period, it begins to fail gradually. The worst part of a tree is the sap-wood, which is next the bark. It is softer than the other parts of the wood, and is liable to premature decay. The deleterious component of the sap-wood is absorbed, if the tree is allowed to grow for a longer period, and in time the old sap-wood becomes proper timber fibre similar to heart-wood. Hence, the goodness of a tree, for timber purposes, depends on the age at which the tree was cut down. When young, the heart-wood is the best ; at maturity, with the exception of the sap-wood, the trunk is equally good throughout ; and when the tree is allowed to grow too long, the heart-wood is the first to show symptoms of weakness, and deteriorates gradually.

The best timber is secured by felling the tree at the age of maturity, which depends on its nature as well as on the soil and climate. The ash, beech, elm, and fir are generally considered at their best when of 70 or 80 years' growth, and the oak is seldom at its best in less time than 100 years, but much depends on surrounding circumstances.

Timber. 109

As a rule, trees should not be cut before arriving at maturity, because there is then too much sap-wood, and the durability of the timber is much inferior to that of trees felled after they have arrived at their full development.

The strength of many woods is nearly doubled by the process of seasoning, hence it is very thriftless to use timber in a green state, as it is not only weak, but is exposed to continual change of bulk, form, and stability. After timber is cut, and before it is properly seasoned, the outside is found to crack, and to split more than the inside of the mass, because it is more exposed to the dessicating effect of the surrounding atmosphere, but as the outside dries, the air gradually finds its way to the interior. If timber is cut up by the saw when green and allowed to season or dry in a gradual manner, it is found to be the most durable. In the arts, however, artificial drying is often resorted to, as in the case of gun stocks. These are put into a dessicating chamber, where a current of air at 90° or 100° is passed over them, at such a rate as to change the whole volume of air in the chamber every three minutes, and it is found that a year of seasoning may thus be saved. The walnut wood is as good after this process, as if the seasoning had been accomplished· by time and exposure, and works more smoothly under the cutting instruments of the stock machinery.

Wood will always warp after a fresh surface has been exposed, and will likewise change its form by the presence of any moisture, either from that contained in the atmosphere or from wetting the surface. The effect of moisture on dry wood is to cause the tubular fibres to swell, hence it is that if a plank or board is wetted upon one side, the fibres there will be distended, and the plank in consequence must bend.

The natural law that governs the shrinking or contraction of timber is most important to practical men, but it is too often overlooked.

The amount of the shrinkage of timber in length, when

seasoning, is so inconsiderable that it may in practice be disregarded. But the shrinkage in transverse directions is much greater, and presents some peculiarities which can only be explained by examining the structure of the wood, as resulting from its mode of growth. An examination of the end section of any exogenous tree, such as the beech or oak, will show the general arrangement of its structure. It consists of a mass of longitudinal fibrous tubes, arranged in irregular circles, which are bound together by means of radial plates or rays, which have been variously named; they are the 'silver grain' of the carpenter, or the 'medullary rays' of the botanist, and are in reality the same in their nature as the pith. The radial direction of these plates or rays, and the longitudinal disposition of the woody fibre, must be considered, in order to understand the action of seasoning. For the lateral contraction or collapsing of the longitudinal fibrous or tubular part of the structure cannot take place without first tearing the medullary rays, hence the shrinking of the woody bundles finds relief by splitting the timber in radial lines from the centre parallel with the medullary rays, thereby enabling the tree to maintain its full diameter. If the entire mass of tubular fibre composing the tree were to contract bodily, then the medullary rays would of necessity have to be crushed in the radial direction to enable it to take place, and the timber would thus be as much injured in proportion as would be the case in crushing the wood in a longitudinal direction.

If an oak or beech tree is cut into four quarters, by passing the saw twice through the centre at right angles, before the splitting and contracting has commenced, the lines $a\,c$ and $b\,c$ in Fig. 10 would be of the same length, and at right angles to each other, or, in the technical language of the workshop, they would be square, but after being stored in a dry place, say for a year, a great change will be found to have taken place, both in the form and in some of the dimensions. The lines $a\,c$ and $b\,c$ will still be of the same

Timber. 111

length as before, but from *a* to *b* the wood will have contracted very considerably, and the two lines *a c* and *b c* will not be at right angles to each other, the angle being diminished, by the portion shown in black in Fig. 10. The

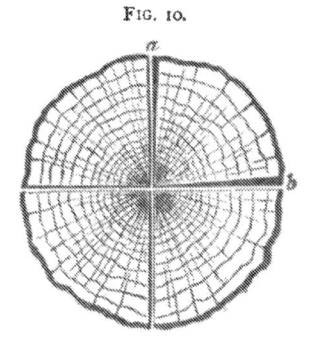

FIG. 10.

medullary rays are thus brought closer by the collapsing of the vertical fibres.

But, supposing that six parallel saw cuts are passed through the tree, so as to form it into seven planks, what will be the behaviour of the several planks? Consider the centre plank first. After due seasoning and contracting, it will be found that the middle of the board still retains the original thickness, from the resistance of the medullary rays, while the thickness will be gradually reduced towards the edges for want of support, and the entire breadth of the plank will be the same as it was at first, for the foregoing reasons, and as shown in Fig. 11. Then, taking the planks at each edge of the centre, by the same law their change and behaviour will be quite different; they will still retain their original thickness at the centre, but will be a little reduced on each edge throughout, but the side next to the heart of the tree will be pulled round or bent convex, while the outside will be the reverse, or hollow, and the plank will be considerably narrower throughout its entire length, more especially on the surface of the hollow side. Selecting the next two planks, they will be found to have

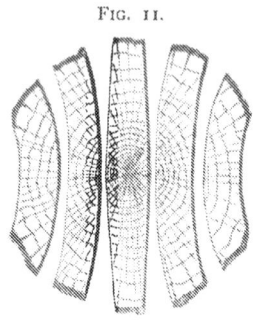

FIG. 11.

lost none of their thickness at the centre, and very little of their thickness at the edges, but very much of their breadth as planks, and will be curved round on the heart side and made hollow on the outside. Supposing some of these planks to be cut up into square prisms when in the green state, the shape that these prisms will assume after a period of seasoning will entirely depend on the part of the tree to which they belonged, the greatest alteration would be

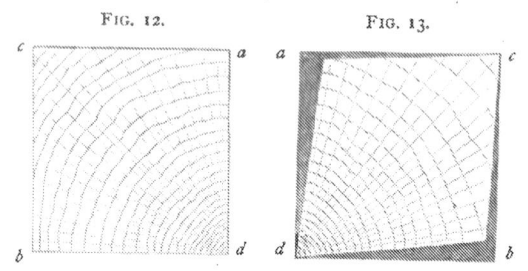

FIG. 12.   FIG. 13.

perpendicular to the medullary rays. Thus, if the square was originally near the outside, as seen in Fig. 12, then the

Timber.

effect will be as shown in Fig. 13, namely, contraction in the direction from *a* to *b*. After a year or two the square end of the prism will become rhomboidal, the distance between *c* and *d* being nearly the same as at first, but the other two edges brought closer together by the amount of their contraction. By understanding this natural law, it is comparatively easy to predict the future behaviour of a board or plank by carefully examining the end wood, in order to ascertain the part of the log from which it has been cut, as the angle of the ring growths and the medullary rays will show this, as in Figs. 14 and 15. If a plank has the appearance of the

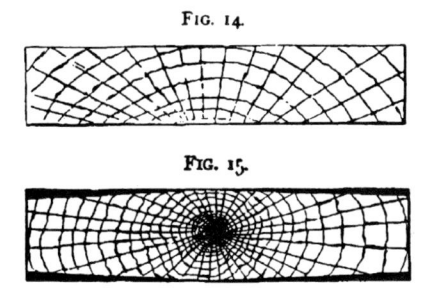

FIG. 14.

FIG. 15.

former, it must have been cut from the outside, and for many years it will gradually shrink in the breadth; while the next plank, shown in Fig. 15, must have been derived from near the centre or heart of the tree, and it will not shrink in the breadth but in thickness, with the full dimension in the middle, but tapering to the edges.

The foregoing remarks apply more especially to the stronger exogenous woods, such as beech, oak, and the stronger home firs. The softer woods, such as yellow Canadian pine, are governed by the same law; but, in virtue of their softness, another law comes into force, which to some degree affects their behaviour, as the contracting power of the tubular wood has sufficient strength to crush the softer medullary rays to some extent, and hence the primary law is so far modified. But even with the softer woods, such as are commonly used in the construction of houses, if the law is carefully observed, the greater part of the evils of shrinking would be obviated. Hence also, it is, that when a round block, as a mast, is formed out of a tree, it retains its roundness because it contracts uniformly or nearly so, whereas if a round spar is formed out of a quartering of the same tree it will become an oval, or otherwise contorted towards that shape.

It would not be in accordance with the object of this book to enumerate all the woods that are employed in the arts, therefore a few only are selected, or such as are commonly employed in the United Kingdom for purposes where strength is the primary object, viz. ash, beech, elm, fir, hornbeam, mahogany, oak, and teak.

Ash.

Ash is a coarse wood, but possessed of considerable strength, and is distinguished for its great toughness and elasticity, and is usually employed where severe shocks and wrenches have to be encountered, such as for agricultural implements, the felloes and spokes of wheels, and the shafts of carriages, for hammer shafts, and for spring purposes generally wherever wood is employed for that purpose.

From its great flexibility it is seldom employed where rigidity is a desideratum. The combination of strength with flexibility is the characteristic of ash, and when the wood is from a young tree, or a tree not too old, it is an invaluable wood in many respects; but as the tree becomes older, the change to brittleness sets in and soon renders it less valuable. It is also remarkable for its endurance when kept dry, but when exposed to damp or to wet it rapidly decays. The numerical value of its properties, as will be seen by the Tables, varies considerably, but in general terms it may be stated that, as compared with oak, good ash has

Timber.

frequently a still greater tenacity and likewise a greater degree of toughness, but from its flexibility, especially when young, it has considerably less stiffness, which unfits it for many purposes.

Beech.

Beech has frequently considerable strength, and is chiefly distinguished for its uniformity, its smoothness of surface, and closeness of grain. It likewise possesses no little beauty, and takes a good polish, more especially when its silver grain is skilfully exposed. When well seasoned and not too old it is frequently used for the cogs of mill gearing, and is usually considered by millwrights as next to hornbeam, both in strength, toughness and general suitability for that purpose. It requires, however, to be kept very dry, for in damp situations it quickly wears out, but, when beech is immersed in water constantly, its endurance is considerable. The strength of beech is nearly the same as that of oak; it is also tougher, but its stiffness is inferior to that of oak, even to the extent of 25 per cent.

Elm.

Elm, although a cross-grained, rough wood, and mostly used for rough purposes, is yet held in great estimation for its toughness and non-liability to split by the driving of bolts. It is much used in the construction of blocks for pulley-tackle, for heavy naval gun-carriages, and for the naves of carriage-wheels. It is a wood which is little affected by constant immersion in water, but decays rapidly when alternately wet and dry, and consequently is not very durable for purposes involving exposure to a wet climate. Its chief defect in ordinary use is its great liability to warp and twist and get out of form; and, as regards strength, toughness, and rigidity, it is inferior to oak, as well as in almost every other respect.

Pine and Fir Wood.

The fir and pine woods are members of a large family, and are of great variety, and differ much in most of their properties. These classes of timber, in addition to being employed for building purposes, are, likewise the chief materials that are used in great works, where the question of strength combined with cost becomes the most prominent consideration. The most durable varieties are the larch, the pitch-pine, and the firs from Memel and Norway, and are valued mostly on account of the large quantity of resin, pitch, and turpentine which they contain. The Canadian pine, variously termed white or yellow, is not a strong wood, but it is much used by engineers for making patterns or models, on account of its smoothness of surface, its non-liability to warp, its comparative freedom from knots, and the facility with which it can be cut. The white or yellow pine is not nearly so strong or so stiff as oak, yet sometimes it is almost equal to it in its tenacity and toughness.

In such a large family as that of the resinous firs and pines, there is almost an equal variation in their strength, toughness, and rigidity. Much of the European wood, such as that from Memel and from some parts of Scotland, is often superior to oak, while at other times it is inferior, more especially in regard to stiffness, as will be seen by the Table.

Hornbeam.

Hornbeam is a wood which is comparatively little used, except by engineers, for the teeth or cogs of wheels and for mallets, for which purposes it is perhaps superior to all other woods, and this is mostly due to its great toughness and remarkably stringy coherence of fibre. Its cohesive strength and other properties depend much upon its age as a plank, and still more on the age of the tree from which the plank was taken. When in the most favourable condition, it is fully equal to the average of oak (even when considered merely as a wood), but when cut from older trees, and when over-seasoned, it is frequently found worthless, and has soon to be renewed. When of proper age and quality, it has no equal for its own special purposes.

Mahogany.

Mahogany is a beautiful close-grained wood, but is used not so much on account of its strength, but more frequently because of its non-liability to shrink, warp, or twist, and from the peculiar property of taking a firm hold of glue. In the last respect it is superior to any other wood. Mahogany differs greatly in regard to its closeness, hardness, strength, and beauty. That from Honduras, called 'bay-wood,' is much inferior to that called 'Spanish' mahogany, which comes from the West Indies; the former is much used in the construction of light textile machinery, but chiefly on account of its cheapness, and the latter is used for furniture or for other ornamental purposes. As regards strength, this wood is inferior to oak in all respects, and its great characteristic defect is unsuitability for exposure to the weather, or indeed for any purpose where it is made alternately wet and dry. When so subjected, it rapidly decays, and loses all its good qualities.

Oak.

Oak, taken as a whole, is one of the strongest and most durable of woods, and is especially adapted for exposure to the weather of a damp climate, and is indeed suitable for almost every purpose where the properties of strength, stiffness, and toughness, combined with endurance, are required. Its value for ship-building is proverbial, and its employment for the staves of casks, for trenails, for carriage-wheels, and for all such purposes requiring lightness and strength in combination it is equally useful. From time immemorial it was esteemed the best timber for heavy roofs, and the condition in which some of these grand old roofs have reached our era, fully attests the wisdom of the selection.

Oak is found of many degrees of quality, but probably none, taking every property into account, is superior to that which grows in England, and which is perhaps more durable than any other. Some of the foreign oaks are as good in some respects, but, as a whole, English is the best.

Teak.

Teak is a most valuable wood, and is fitted for most purposes where oak is employed. It is equally durable, but is more liable to split by the driving of bolts, and is not so well adapted for a sudden jar or wrench. Its great durability is partly due to the large quantity of oil which it contains ; the oil likewise is found to prevent the rusting of the iron bolts employed in framing it. For gun-carriages or for similar purposes, its value is increased by the small amount of shrinkage which takes place, even after long exposure to a hot climate. In this respect it is superior to oak ; and it has another property, it is not much exposed to attack by the worm of tropical countries, which is a great advantage. Teak, as a rule, is superior to oak, both in regard to tenacity and stiffness, but is inferior as regards toughness ; but when those qualifications are required in combination with endurance in a tropical country, it has the superiority, and is therefore preferred.

The foregoing remarks being of a general nature, can only be considered as preliminary to a study of the Tables which follow. In past times, an immense number of experiments have been made upon wood, more especially in regard to its tensile strength. The next Table contains the result of experiments made at different times to ascertain the general range of the ultimate cohesion or tensile strength of the foregoing descriptions of timber.

TABLE XVIII.

These results have been collected from various sources, but chiefly from the experiments made by Barlow, Bevan, and Muschenbroek.

Timber.	Ultimate tenacity in lbs. per square inch of section.
Ash	from 19,600 to 15,784
Beech	,, 22,200 ,, 11,500
Elm	,, 14,400 ,, 13,489
Fir	,, 18,100 ,, 7,000
,, American	12,000
,, Memel	11,000
,, Riga	12,600
,, Mar Forest	12,000
,, Larch	10,200
Hornbeam	from 20,240 to 4,253
Mahogany	,, 21,800 ,, 8,000
Oak	,, 19,800 ,, 9,000
,, English	15,000
,, African	14,400
,, Canadian	12,000
,, Dantzic	14,500
Teak	from 15,000 to 8,200

The next Table contains the result of experiments made to ascertain the average resistance to ultimate crushing of the foregoing descriptions of timber in the direction of the fibre, when in the form of short pillars.

TABLE XIX.

These results are mostly derived from the experiments of Hodgkinson.

Timber.	Average resistance to crushing stress, in lbs. per square inch of section.
Ash	from 9,363 to 8,683
Beech	,, 9,363 ,, 7,733
Elm	10,331
Fir	from 6,819 to 5,375
Hornbeam	,, 7,289 ,, 4,533
Mahogany	8,280
Oak, English	from 10,058 to 6,484
,, Dantzic	7,731
Teak	12,101

This Table contains the result of experiments made by Rennie, to ascertain the average resistance to crushing in the direction of the fibre of short pillars of timber.

TABLE XX.

Size of specimens, 1-inch cubes.

Timber.	Average resistance to crushing stress, in lbs. per square inch of section.
Elm	1,284
American pine	1,606
White deal	1,928
English oak	3,860

These experiments show a considerably less result than those made by Hodgkinson, and it is scarcely possible to reconcile the discrepancy between the two Tables; but most probably it could be easily explained if all the facts of the experiments were fully known, as everything depends on the amount or degree of the *crushing* action that was effected in the several experiments.

Timber is sometimes subjected to a crushing force in the other direction; namely, at right angles to the fibre, as in the case of a wedge, or when baulks of timber are used to support great weights, or when two pieces are framed and bolted together or otherwise. In regard to this important point, however, there are comparatively few experiments recorded, and, as a rule, those which are recorded are not very definite. All are familiar with the fact, that a very small force will be sufficient to indent wood to some appreciable extent. The precise force that will crush a cubic inch perceptibly—say $\frac{3}{1000}$ths of an inch—may be said to range from 500 lbs. to 1,500 lbs., according to the hardness or softness of the specimen.

The following Table contains the results of experiments made by Hatfield, to ascertain the force required to compress timber to a depth of $\frac{1}{20}$th of an inch, in a direction at right angles to the fibre :—

TABLE XXI.

Timber.	Stress in lbs. per square inch required to crush fibres transversely, and to produce an indentation of $\frac{1}{20}$ of an inch.	Specific gravity.
Spruce fir	500	·369
White pine	600	·388
Mahogany, Bay-wood	1,300	·439
,, St. Domingo	4,300	·837
Oak	1,900	·612
Ash	2,300	·517
Lignum vitæ	5,800	1·282

On the Strength of Materials.

Shearing or Detrusion of Wood.

In the application of timber to form framings for machinery, in the construction of roofs of houses, and in scaffoldings, too much reliance is frequently placed by the carpenter, upon the support which is understood to be derived from an abutment against a notch cut into the solid wood, thereby bringing into play the end shear of a portion of the material, in the direction of the fibre. From experiments made by Barlow, the strength of timber thus strained is only about $\frac{1}{20}$th of the tenacity of the same wood in the direction of its length.

The assumption that wood may be treated like metal in this respect is wrong. With metals, the resistance to tension and detrusion are nearly alike both ways, the solid 'joggles' left upon an iron casting give an abutment nearly equal to the tensile strength of the metal; but with wood similarly arranged it is only equal to the $\frac{1}{20}$th part of the tensile strength. This deficiency is usually compensated for by the extra length given to the part which has to be detruded; still, the weakness of timber in this respect should be noted.

The following Table refers to the transverse strength of timber when used as a beam. The results are collected from a variety of sources; and as the experiments were made with specimens of varied dimensions, the whole have been reduced to one standard, for the sake of comparison :—

TABLE XXII.
Results of Experiments made to ascertain the Transverse Strength of Timber.

The experimental beams were of various dimensions, and were also supported and loaded in different ways, their actual breaking weights are shown in column 4, and the numbers in column 5 are the calculated weights required to fracture beams of the same timber, 1 foot long and 1 inch square, when supported at each end and loaded at the centre.

1	2	3	4	5	6	7
Timber.	The experimental beams. Dimensions.	Position of the load and the supports.	Breaking weight in lbs	Calculated breaking weight in lbs. of a beam 1 foot long and 1 inch square.	Authority.	Remarks.
Ash	7′ long × 2″ square	Supported at both ends and loaded at the centre.	772	675	P. Barlow.	
	4′ 2″ ,, × 2″ ,,		1,304	679	P. W. Barlow.	
	2′ 6″ ,, × 1″ ,,		324	810	Tredgold.	Dry.
	2′ 6″ ,, × 1″ ,,		254	635	Ebbels.	From a young tree.
	2′ ,, × 1″ ,,		314	785	Tredgold.	Medium quality.
	2′ ,, × 2″ deep and 1″ wide	Supported at one end and loaded at the other.	326	652	P. Barlow.	
	5′ ,, × 2″ square		239	595	Peake & Barrallier.	
Beech	7′ long × 2″ square	Supported at both ends and loaded at the centre.	593	518	P. Barlow.	Medium quality.
	2′ 6″ ,, × 1″ ,,		271	677	Ebbels.	
	2′ ,, × 2″ deep × 1″ wide	Supported at one end and loaded at the other.	352	704	P. Barlow.	
	3′ ,, × 2″ square		401	602		

TABLE XXII.—(continued.)

The experimental beams.

Timber	Dimension	Position of the load and the supports.	Breaking weight in lbs.	Calculated breaking weight in lbs. of a beam 1 foot long and 1 inch square.	Authority.	Remarks.
Elm	7' long × 2" square 4' 2" ,, × 2" ,, 4' 2" ,, × 2" ,, 2' 6" ,, × 1" ,,	Supported at both ends and loaded at the centre.	386 772 660 216	337 402 343 540	P. Barlow. P. W. Barlow. Ebbels.	Dry, and both from the same tree.
Fir (American)	2' long × 1' square 4' ,, × 3" ,, 2' ,, × 3" ,,	Supported at both ends and loaded at the centre. Supported at one end and loaded at the other.	285 3,259 1,753	570 483 519	Tredgold. Fincham.	White spruce. Red pine.
Fir (Christiana)	2' long × 1" square 4' 2" ,, × 2" ,, 4' 2" ,, × 2" ,,	Supported at both ends and loaded at the centre.	343 940 1,052	686 489 548	Tredgold. P. W. Barlow.	Dry, and from the same deal.
Fir (Larch)	7' long × 2" square 2' 6" ,, × 1" ,, 2' 6" ,, × 1" ,, 2' 6" ,, × 1" ,, 5' ,, × 2" ,,	Supported at both ends and loaded at the centre. Supported at one end and loaded at the other.	325 253 223 129 162	284 632 557 322 405	P. Barlow. Tredgold. Peake & Barrallier.	Choice specimen. Medium quality. Very young wood. Dry.

Timber. 125

Fir (Mar Forest)	7′ long × 2″ square	Supported at both ends and loaded at the centre.	436	381	P. Barlow.
Fir (Memel)	4′ 2″ long × 2″ square 2′ 6″ ,, × 1″ ,,	Supported at both ends and loaded at the centre.	1,108 218	577 545	P. W. Barlow. Tredgold.	Dry.
Fir (Riga)	7′ long × 2″ square 2′ 6″ ,, × 1″ ,, 4′ ,, × 2″ ,, 5′ ,, × 2″ ,, 2′ ,, × 3″ ,,	Supported at both ends and loaded at the centre. Supported at one end and loaded at the other.	422 212 210 153 1,974	369 530 420 382 584	P. Barlow. Tredgold. Beaufoy. Peake & Barrallier. Fincham.	Dry. Dry.
Oak (Adriatic)	7′ long × 2″ square	Supported at both ends and loaded at the centre.	526	460	P. Barlow.
Oak (African)	4′ 2″ long × 2″ square	Supported at both ends and loaded at the centre.	1,643	855	P. W. Barlow.	Cut from very fine seasoned timber.
Oak (Canadian)	4′ long × 3″ square 7′ ,, × 2″ ,,	Supported at both ends and loaded at the centre.	3,863 673	572 589	Fincham. P. Barlow.
Oak (Dantzic)	7′ long × 2″ square 4′ ,, × 3″ ,, 4′ ,, × 2″ ,,	Supported at both ends and loaded at the centre. Supported at one end and loaded at the other.	560 4,450 196	490 659 392	P. Barlow. Fincham. Beaufoy.

TABLE XXII.—(continued.)

Timber.	The experimental beams. Dimensions.	Position of the load and the supports.	Breaking weight in lbs.	Calculated breaking weight in lbs. of a beam 1 foot long and 1 inch square.	Authority.	Remarks.
Oak (English)	7′ long × 2″ square 2′ ,, × 1″ ,, 2′ 6″ ,, × 1″ ,, 2′ 6″ ,, × 1″ ,, 4′ 2″ ,, × 2″ ,,	Supported at both ends and loaded at the centre.	637 482 218 284 999	557 964 436 710 520	P. Barlow. Tredgold. Ebbels. P. W. Barlow.	From a young tree. From an old tree. Medium quality. Fast grown.
	4′ ,, × 2″ ,,	Supported at one end and loaded at the other.	266 210	532 420	Beaufoy.	
Mahogany (Honduras) Mahogany (Spanish) Mahogany (New South Wales)	2′ 6″ long × 1″ square 2′ 6″ ,, × 1″ ,, 4′ ,, × 3″ ,,	Supported at both ends and loaded at the centre.	255 170 4,119	637 425 610	Tredgold. Fincham.	Seasoned wood.
Teak.	7′ long × 2″ square 4′ ,, × 3″ ,, 3′ ,, × 1¼″ ,,	Supported at both ends and loaded at the centre.	938 4,897 1,213	820 721 1,075	P. Barlow. Fincham. Mayne.	Malabar. Johore.
	5′ ,, × 2″ ,,	Supported at one end and loaded at the other.	257	642	Peake & Barralier.	Old and dry.

Timber.

The attention of the student is particularly directed to column 5 of the foregoing Table, which will furnish him with much useful knowledge on the transverse strength of wooden beams, in a small compass, and will enable him to calculate either the strength or required dimensions of any other rectangular beam of any of the kinds of wood therein specified.

In Chapter XII., upon the strength of structures, the principles which regulate the strength of such beams are fully stated. It will be sufficient here to remark, that by taking the beam in column 5, which is 1 foot long and 1 inch square, as a standard, the student can ascertain the strength of any other rectangular beam, by simply multiplying the strength of the standard beam by the breadth, and the depth squared, both in inches, and then dividing the product by the length, in feet, of the given beam, the strength of which he wishes to ascertain.

Table showing approximately the mean breaking weight of beams of timber, 1 foot long and 1 inch square, supported at both ends and loaded at the centre, deduced from the experiments enumerated in Table XXII. :—

TABLE XXIII.

Timber.	Mean breaking weight of beam in lbs.	Timber.	Mean breaking weight of beam in lbs.
Ash . . .	690	Oak—	
Beech . . .	625	,, Adriatic .	460
Elm . . .	405	,, African .	855
Fir—American .	524	,, Canadian .	580
,, Christiana .	574	,, Dantzic .	513
,, Larch .	440	,, English .	591
,, Mar forest .	381	Mahogany .	557
,, Memel .	561	Teak . . .	814
,, Riga .	457		

The above Table can only be considered as available for approximate calculations.

CHAPTER IX.

TRANSVERSE STRENGTH OF IRON AND RESISTANCE TO IMPACT.

THE transverse strength of materials, more especially that of cast iron, in the form of beams and girders, has been closely investigated by a number of scientific men, particularly by Sir William Fairbairn, Mr. Hodgkinson, and the Railway Commissioners. The subject will be treated of in Chapter XII. ; but it will be convenient to give here an abstract of the results of various experiments, on the transverse fracture of bars of different materials.

A bar laid horizontally upon supports and loaded at the centre bends, and ultimately breaks, by transverse strain. The two points most important to be observed are the breaking weight and the deflection of the bar. Some results, obtained in experiments of this kind, are given in the subsequent Tables.

'The following Table contains a synopsis of the results of a series of experiments, carried out under the direction of the Commissioners appointed to enquire into the Application of Iron to Railway Structures. Their object was to determine the transverse strength and ultimate deflection of cast-iron bars, when subjected to a statical load placed at the centre :—

TABLE XXIV.

Name of Iron.	Size of bar in inches.	Distance between supports in feet.	Weight of bar between supports in lbs.	Breaking weight in lbs.	Ultimate deflection in inches.
Blaenavon No. 2.	1 inch square.	4½	13·49 13·34 13·02 13·06	461 359 437 423	1·796 1·315 1·85 1·6917
Mr. Stirling's Calder No. 1, with 20 per cent. of malleable iron scrap.	2 inches square.	9	111·07 114·41 111·46 112·91	2,411 1,837 2,083 2,364	3·103 2·227 2·395 2·883
Blaenavon No. 2.	2 inches square.	9	107·6 109·6 108·0 108·0	1,249 1,414 1,121 1,097	2·906 3·486 2·527 2·498
Blaenavon No. 2.	3 inches square.	13½	358·29 365·39 367·29 365·06	2,698 2,671 3,389 2,686	4·863 4·3908 5·024 4·391

Abstract of the results of experiments made by the above Commissioners to find the relative transverse strength of cast-iron bars, of various sizes, but similarly proportioned in all their dimensions :—

Note.—The horizontal pressures in the following Table were computed from the vertical pressures, by taking into account the weight of the bar itself, which would not have any tendency to weaken the bar when pressed in the horizontal direction, but of course acted as so much additional load when the bars were pressed vertically.

TABLE XXV.

Size of bar.	Vertical pressures.		Horizontal pressures computed from the vertical pressures.	
	Mean strength in lbs.	Mean ultimate deflection in inches.	Mean strength in lbs.	Mean ultimate deflection in inches.
3 bars 4½ feet span and 1 inch square.	440	1·779	447	1·808
6 bars 9 feet span and 2 inches square.	1,338	3·0035	1,394	3·126
4 bars 13½ feet span and 3 inches square.	2,861	4·667	3,043	4·966
3 bars 6¾ feet span and 3 inches square.	6,117	1·2916	6,207	1·311

This Table shows, first, that the strength of the similar bars, 1 inch, 2 inches, and 3 inches square, and 4½ feet, 9 feet, and 13½ feet span, to resist a horizontal pressure, are respectively 447, 1,394, and 3043 lbs.; but if the elasticity of the bars had been perfect, their strengths should have been as the squares of their linear dimensions, namely, in the ratio of the numbers 1, 4, and 9. Dividing the strengths by these numbers, the quotients ought, on the supposition of perfect elasticity, to be equal. We find, however,

$$447 \div 1 = 447 \text{ for 1 inch bars.}$$
$$1,394 \div 4 = 349 \text{ for 2 inch bars.}$$
$$3,943 \div 9 = 338 \text{ for 3 inch bars.}$$

The quotients being unequal, and the greatest deviation being in the case of the smallest bar. Probably this is due

Transverse Strength & Resistance to Impact. 131

to the more rapid cooling of the small bars, which tends to increase the strength and hardness of the metal. The Table shows, also, that for square bars of constant length between the supports, the transverse strength varies nearly as the cube of the side of the square.

Abstract of experiments made for the Railway Commissioners on the resistance of cast-iron bars to long-continued impact, from a ball striking horizontally against the middle of the bar:—

TABLE XXVI.

Distance between supports.	Side of square of bar inches.	Weight of ball in lbs.	Velocity of impact in feet per second.	Assigned deflection in inches.	Number of blows.	Effect.	Remarks.
13¼ ft.	3	151¼	8·812	⅛ or 1·5	1,085	broken.	slightly defective.
	3	151¼	9·4899	⅛ or 1·5	4,000	not broken.	
	3	603	3·6156	⅛ or 1·5	4,000	,,	
	3	603	4·5754	5/12 or 1·875	1,350	broken.	slightly defective.
	3	151¼	13·7834	¼ or 2·25	127	,,	,,
	3	603	5·0839	½ or 2·25	3,026	,,	
9 ft.	2	75½	6·2132	⅛ or 1·00	4,000	not broken.	
	2	603	1·8071	⅛ or 1·00	4,000	,,	
	2	75½	—	⅛ or 1·50	229	broken.	defective.
	2	75½	9·3204	⅛ or 1·50	1,282	,,	slightly defective.
	2	603	2·5284	⅛ or 1·50	3,695	,,	,,
	2	75½	11·693	⅜ or 2·00	127	,,	
	2	603	3·5872	⅜ or 2·00	474	,,	
4¾ ft.	1	75½	2·1461	⅛ or ·50	4,000	not broken.	
	1	75½	3·0368	⅛ or ·75	4,000	,,	
	1	75½	3·5018	1/12 or ·875	3,700	broken.	slightly defective.

One bar only, and that a small one, stood 4,000 blows, each blow bending it through half its ultimate deflection with a statical breaking weight; but all the bars, when sound, stood that number of blows, each blow deflecting them through ⅓rd of their ultimate deflection. A cast-iron bar will, however, be bent through ⅓rd of its ultimate deflection with less than ⅓rd of its breaking weight, laid on gradually, or ⅛th of the breaking weight laid on at once—hence the prudence of making beams capable of bearing more than six times the greatest weight which will be laid upon them.

Abstract of results of experiments on vertical impacts, upon unloaded and loaded beams of cast iron (Blaenavon, No. 2), 13 ft. 6 ins. between supports and 3 ins. square; the falling weight being 303 lbs. :—

TABLE XXVII.

Weight of beam in lbs.	Additional load on beam in lbs.	Height of fall necessary to break the beam in inches.	Velocity of impact due to that height.
376¾	nil.	28½	
382	4 lbs. at centre.	33	13·3
375½	28 lbs. at centre.	42	15·0
387	166 lbs. spread over the beam + 4 lbs. at centre.	48	16·0
386	389¼ lbs. spread over the beam + 4 lbs. at centre.	48	16·0
379	391 lbs. spread over the beam + 4 lbs. at centre.	66	18·8
382	956¼ lbs. spread over the beam + 4 lbs. at centre.	60	17·9

It will be seen that loaded beams resisted better than unloaded ones.

The deflections were found to vary nearly as the velocity of impact.

The sets were very great, but did not appear to injure the strength of the bars more than in ordinary cases, with an unloaded beam.

Transverse Strength & Resistance to Impact. 133

Synopsis of experiments on the transverse flexure of five bars of Blaenavon iron, No. 2, cast to be $3\frac{1}{2}$ inches × $1\frac{1}{2}$ inch in section, and 13 feet 6 inches between supports, the pressure being applied in the direction of their least dimension :—

TABLE XXVIII.

Weight applied acting horizontally in lbs.	Mean deflection after 5 minutes in inches.	Mean set after 5 minutes in inches.	Remarks.
28	·181	·0016	
56	·3754	·0068	
112	·7686	·0192	
168	1·184	·0468	
224	1·632	·0914	
280	2·105	·1486	Limit of elasticity,
336	2·604	·2266	$\frac{1}{30}$ of ultimate or
392	3·169	·3292	breaking weight.
448	3·756	·4574	
504	4·402	·6078	
560	5·035	·7854	
616	5·777	1·038	
672	6·565	1·287	
728	7·610	1·707	
784	8·73	2·186	
840	9·887	2·691	Nos. 1, 3, 4, 5, broke.
896	10·7 (No. 2)	3·006	
934	—	No. 2 broke.	

Similar experiment with bars 2 inches × 1 inch, and 9 feet between supports :—

TABLE XXIX.

Weight applied acting horizontally in lbs.	Mean deflection after 5 minutes in inches.	Mean set after 5 minutes in inches.	Remarks.
28	·235	·002	Limit of elasticity, $\frac{1}{17}$ of ultimate or breaking strength.
485	mean breaking weight of the sound bars.		

134   *On the Strength of Materials.*

These Tables show the apparently very great want of elasticity of cast iron: a permanent set being attained in one case with $\frac{1}{30}$th of the ultimate breaking weight, and in another case with $\frac{1}{17}$th of the ultimate breaking weight.

The following results are intended to show the effect produced by repeated deflection—the experiments having been made for the Railway Commissioners.

Cam experiments on the deflection of cast iron bars, 13 feet 6 inches long, and 3 inches square :—

The *step cam* gradually deflected the bars to the required amount, and then allowed them to spring back instantly to their original position, so far as they would do so.

The *rough cam* was simply a toothed excentric, working into a rack fixed upon the bar, it deflected and also allowed the bar to return to its original position, gradually, and imparted, by its roughness, a highly vibratory motion to the bar.

First Experiment.—Three bars of No. 2 Blaenavon iron were subjected to 10,000 deflections by the *rough cam*, each deflection amounting to that caused by $\frac{1}{3}$rd of the statical breaking weight. The permanent set at 100 deflections was found to be ·2 inch, ·17 inch, and ·19 inch respectively, and there was no increase in the permanent set after 100 deflections. These bars required the same weight to break them after the experiment, as similar bars which had not been so treated, showing that they were uninjured by the experiment.

Second Experiment.—This was similar to the first experiment, but the deflections were made with a *step cam*. The permanent sets at 150 deflections were ·16 inch, ·13 inch, and ·12 inch. No increase was observed after 150 deflections, and the bars were apparently not weakened by the experiment.

Third Experiment.—Similar bars to the above were deflected to the amount caused by $\frac{1}{3}$rd of the statical breaking weight, by the *step cam*.

Transverse Strength & Resistance to Impact.

No. 1 bar broke after 51,538 deflections—(fracture good) —bar weakened by experiment.

No. 2 bar broke after 25,486 deflections—(flaw).

No. 3 bar not broken after 100,000 deflections—bar not weakened by experiment.

Permanent set at 150 deflections, ·08 inch, ·2 inch, ·08 respectively—no further increase.

Fourth Experiment.—Two similar bars of Clyde iron, No. 3, deflected to the amount caused by ½ the statical breaking weight by the *rough cam*.

No. 1 bar not broken after 30,000 deflections—(bar not weakened).

No. 2 bar broke after 28,602 deflections.

Permanent set at 1,000 deflections, ·35 inch and ·37 inch respectively—no further increase.

Fifth Experiment.—Three bars of No. 2 Blaenavon iron, deflected to the amount caused by ½ the statical breaking weight by the *step cam*.

No. 1 bar broke after 617 deflections.
No. 2 bar broke after 490 deflections.
No. 3 bar broke after 900 deflections.

Permanent set at 150 deflections, ·44 inch, ·37 inch, and ·37 inch, no further increase.

Sixth Experiment.—Bar of wrought iron, 9 feet long and 2 inches square, showing statical weight required to obtain certain deflections:—

TABLE XXX.

Deflections, inches.	Weights, lbs.	Permanent set.	Remarks.
·333	507	0	After the bar had 1,950 lbs. on, it suddenly gave way; it did not break, but no further weight could be applied with certainty.
·666	926	0	
·833	1,121	0	
1·000	1,364	·054	
1·8	1,950	·86	

Seventh Experiment.—This was a valuable experiment; it showed that a wrought iron bar, 9 feet long and 2 inches square, deflected to ·833 inch (about ⅔ths of the strain that permanently injured a similar bar), withstood 100,000 deflections: permanent set ·15 inch; bar uninjured.

Eighth Experiment.—Five similar bars, deflected by the step cam:—

TABLE XXXI.

No. of bar.	Amount of deflection.	Number of deflections.	Permanent set.
	inches.		
1	·33	10·000	0
2	·66	10·000	0
3	·83	10·000	0
4	1·00	10·000	·06
5	2·00	10	·3
		50	·54
		100	·69
		150	·84
		200	·98
		300	1·84

CHAPTER X.

RESISTANCE TO TORSION AND SHEARING.

THE torsional resistance of different materials has not received so much investigation as the tenacity, compressive resistance, and transverse resistance of materials, and the results of the few recorded experiments are very discordant. Still there are sufficient data to show the relative value of the different metals, and their approximate resistance to torsion.

It is chiefly in designing the shafting of mills and factories that we require to know the torsional strength and stiffness of bars. Such shafting consists almost always of cylindrical bars of greater or less length, and it is the agent by which the power of the steam-engine or other motor is distributed to the machinery, spread frequently over a large area. But the crank shafts of engines, the propeller shafts of screw steam vessels, the spindles or shafts of cranes, and many other parts of machines are also subjected to torsion.

The most important result of the experiments on torsion, which should be remembered in designing shafting, is this, that the strength of cylindrical bars, when twisted, is proportional to the cubes of their diameters. Hence, if we know the torsional strength of any one cylindrical bar, say of 1 inch in diameter, when made of cast iron, wrought iron, or steel, we can calculate from it the strength of shafts of other dimensions of the same material. If the strength of such a shaft or bar of 1 inch in diameter is represented by 1, that of a shaft 2 inches in diameter will be equal to 8, of 3 inches to 27, and so on. In the case of hollow shafts, by simply cubing the exterior diameter and then deducting the cube of the interior diameter, the difference will give the relative value of the shaft, as an agent to transmit power or motion.

The torsional strength of wrought iron has received more attention than that of any other material, because modern shafting is chiefly made of that metal. We may therefore indicate, first of all, the results obtained in experiments with wrought iron. Many experiments on torsion have been made on cylindrical bars 1 inch in diameter, the load being applied by a lever, the length of which is 12 inches measured from the centre of the bar. We may therefore express the relative resistance to torsion of different bars by stating the weight which twists them asunder when applied in the way indicated. The amount of load which is required to produce rupture is usually reckoned at an average of 1,000 lbs. on the end of the lever, but will necessarily depend on the quality of the iron as well as other conditions. In one instance, where a bar or strip of iron was cut out of a welded coil, across the welds, rupture took place with less than

400 lbs., no doubt in consequence of a defective weld. The highest torsional strength here quoted, namely, 1,000 lbs. acting at a leverage of one foot, is derived from experiments made with specimens of an exceptionally good quality of wrought iron; 700 lbs. to 800 lbs. is probably nearer to the average strength of good common wrought iron.

Wrought iron is better for shafting than cast iron, because of its greater torsional strength. In consequence of the greater strength of wrought iron, shafting made of that material is lighter than cast-iron shafting, and this is a matter of great importance, because the friction of shafts on their bearings is directly proportional to their weight. As will be shown, however, presently, the stiffness of wrought-iron shafting is not much greater than that of cast-iron shafting, and permanent set commences with less stress in the former case than in the latter.

Cast iron, even more than wrought iron, varies in quality. From some experiments made with specimens of cast iron, the torsional resistance was equal to 900 lbs. on the end of the lever; but the highest result obtained at Woolwich, with a specimen which was considered a particularly good sample of cast iron, gave only 670 lbs., and there are some kinds of cast iron which would not afford a resistance equal to the half of even that weight. Hence it may be convenient to assume the torsional strength of a cylinder one inch in diameter of good cast iron as equal to from 650 lbs. to 750 lbs., according to the quality and the conditions of the casting.

Some experiments were made in America with cast iron under torsion, the bar being 8 diameters long. The force required to give to the bar a permanent set of $\frac{1}{2}°$ was equal to $\frac{7}{10}$ths of the force required to twist it asunder; the same amount of set is given to wrought iron with a less stress, or about $\frac{6}{10}$ths of the ultimate twisting force.

From some experiments made by Sir W. Fairbairn with cast iron, some specimens were higher, and some were lower, but the mean was 733 lbs. on the end of the 12-inch lever.

Cast steel, being superior to wrought iron in most other respects, is also stronger in its resistance to torsional stress. From some experiments which have been made with steel of high quality, it was found that a cylinder one inch diameter required 1,900 lbs., acting at a leverage of one foot, to twist it asunder. Such a high result, however, must be rare, and, judging by the result of an experiment made with part of the tempered steel lining of a gun, which only required 1,355 lbs., we may safely consider the value of the average quality of steel, as much less than 1,500 lbs. on the 12-inch lever; that of good Bessemer steel being 1,150 lbs.

Cylinders of wrought copper, one inch in diameter, require a load varying between 400 lbs. and 450 lbs. on the end of the 12-inch lever.

The foregoing remarks refer to the ultimate torsional or breaking strength; but, as before remarked with regard to the other properties, the limit of torsional elasticity is the chief point for the engineer to keep in mind, regarding which, however, there are not sufficient reliable data on which to lay down simple rules. Hence, it is usual not to load shafts with more than $\frac{1}{10}$th of the weight that would be required to produce ultimate rupture; there are some, however, who are satisfied with the half of that margin.

The torsional stiffness of a long line of continuous shafting is dependent on the length; but the length of the line affects the amount of twist only in this way, that the total twist of the entire length is the sum of the twists of each portion of the length. The amount of twisting of a given length is proportional to the twisting force, so long as that force does not exceed the elastic limit of the material. If it does, a permanent twisting is produced, similar to the permanent set in experiments on tension.

In the case of long shafting, the torsional stiffness is of more importance, practically, than the torsional strength. If the shaft is deficient in stiffness, so that when working it is twisted through a large angle, it runs in a jerky manner, and the machines driven by it work badly. In long shafting, it is usual to secure sufficient stiffness by restricting the angle of torsion to some definite limit, say to $\frac{1}{4}°$ per yard length of the shaft, and to secure this amount of stiffness a larger shaft is often required than would be needed, if strength alone were considered. The resistance of shafts of equal stiffness, in this sense, is proportional to the square of the area—that is, a 2-inch shaft will transmit 16 times the force which would be transmitted by a 1-inch shaft without being twisted through a greater angle.

In the shafts or spindles of cranes, torsion, and not stiffness, is the prominent point, on account of their shortness, which does not give sufficient play to elasticity to make the want of stiffness objectionable.

When a tie-rod, or a strut, or a long shaft, is so constructed that certain parts have a larger or smaller diameter than other adjoining parts, it is evident that the strength cannot be greater than that of the weakest portion. In practice, however, it is found that if the small part of the shaft passes abruptly into the larger part, it is even less strong than its dimensions would indicate, or rather, it is more correct to say, that fracture takes place with a less load than in an uniform bar, of the same section as the smallest part of the shaft. More especially is this the case with the bearings of shafts, if the re-entrant angle of journals or corner of union is left square, instead of being rounded off. This discrepancy arises from two obvious causes : first, because the small part is the weakest portion of the bar, and consequently has more than its share of the elastic work to perform ; and secondly, because at the junction of the two diameters, the fibre or molecules do not spread out to take hold of the larger diameter. In practice, it is found that this defect is obviated to some extent by rounding the corner, and the risk of fracture is thus reduced.

In considering the diameter to be given to a shaft, the pull or transverse strain which will come upon it in the

Resistance to Torsion and Shearing. 141

middle of the distance between the bearings must also be taken into account. A comparatively small shaft may be strong enough to transmit the necessary power, yet not strong enough to bear the stress of several tight straps, causing deflection in addition to twisting, and hence the strength of the shaft has so far to be reckoned as a beam, irrespective of torsion, and a suitable diameter provided accordingly.

The results of experiments made with round bars of 1-inch diameter, and with a lever of 12 inches in length, have been given, and it has been stated that the strength of other bars is to that of a 1-inch bar as the cubes of the diameters. It has also to be observed, that the resistance of the shaft to a given load is inversely as the radius of the lever. With a lever 6 inches long, or half the standard length of 12 inches, twice the load must be applied to break the bar. With a lever 24 inches long, or double the standard length, only half the load would be sufficient to break the bar. Keeping this in view, we may compare the torsion produced by wheels or pulleys of different diameters.

In practical operations, it is seldom that the engineer can obey the laws of correct proportion as strictly as he would desire, on account of other considerations, which step in to modify his proceedings. If, for example, a long shaft were to be constructed, of sectional area in exact proportion to the strain anticipated at each point, the shaft ought to be largest at the driving end, and gradually diminished by tapering down to the other extremity. In practice, however, this would be found inconvenient, because the arrangement involves a multifarious assortment of pulleys, with bores of different diameters, for the reason that, in the conduct of a factory, such pulleys have frequently to be transposed from one point of the shaft to another point, in order to suit the varying circumstances of the machinery to be driven. To obviate this difficulty, true proportion in the shaft is constantly departed from, and the shaft usually made of the proper strength at the driving end, to transmit the required amount of power, and then reduced by stages, at, say every hundred feet, or even less. This implies greater cost for shafting in the first instance, but the future facility which it affords is more than an equivalent to the original outlay.

In the working of a long line of shafting, the elasticity is observable, and even becomes inconvenient and prominent when the shaft is made too light. The following interesting experiment was made, in connection with the driving of such a long line. The shaft was originally driven from one end only, and afterwards means were adopted to drive from both ends; but the irregularity of the motion, with the new arrangement, gave rise to a considerable disturbance in the middle of the shaft, which made the new arrangement equally unsatisfactory with the previous one. The shaft was then cut in the middle, and driven at the same velocity from both ends. Long pointers were fastened to each shaft at the point where they came together, in order to show the divergence of their motion; the cause of the previous disturbance was soon made evident, for the two bars, although running at the same speed in revolutions per minute, did not keep true time in their twisting, the two pointers were seldom together, being sometimes to the extent of half a revolution apart, sometimes one, and at other times the other, in advance, the cause of disturbance being the ever-changing strain, due to variations in the stress of the work done at various points; stiffness and steadiness can only be attained by an increase of substance beyond that necessary for strength; the line was evidently deficient in stiffness, notwithstanding its abundance of strength.

When a bar of iron is wrenched asunder by torsion, the break or rupture must be considered as a shear. Where a given area of metal has to be sheared, it is found that the force required is nearly the same as that necessary to tear

Resistance to Torsion and Shearing.

asunder a piece of the same material of equal area. The shearing force, however, is rather less than the tenacity, therefore, to shear a square inch bar would require a force of between 18 and 24 tons, according to quality. The torsional strength of the bar may be calculated from this value of the shearing force, but practically the question does not present itself in this form, and it is more simple to consider torsion in connection with forces acting at the end of a lever.

It will thus be seen that the strength of shafts to resist torsion depends on four conditions : first, on the strength of the material ; second, on the diameter of the shaft ; third, on the length of the lever employed in twisting ; and fourth, upon the force or weight applied at the end of the lever.

The following Table shows the values assigned in the foregoing remarks as the ultimate strength of a bar 1 inch in diameter, the weight being applied to a lever 12 inches long.

TABLE XXXII.

Cast steel	High	1,900 lbs.
	Ordinary.	1,500 ,,
	Mild	1,355 ,,
	Bessemer.	1,150 ,,
Wrought iron		700 to 1,000 ,,
Cast iron		650 to 750 ,,
Wrought copper		400 to 450 ,,

The following Table conveys some precise knowledge in regard to torsion :—

TABLE XXXIII.

Resistance of Metals to Torsion.

Size of specimen.	Nature of material	Length of lever.	Weight on end of lever required to destroy elasticity.	Ultimate breaking weight on end of lever.	Permanent set. 1 turn being = to 1. Weight on end of lever.	Proportion of set.	Ultimate torsion 1 turn being equal to 1.	Authority.
Diameter 1 inch. / Length of twisted part 4 diameters.	Mild homogeneous cast steel tempered in oil.	12 ins.	lbs. 552	lbs. 1,355	lbs. 785	·058	·602	Kircaldy.
	Strong cast iron.		—	670	536	·003	—	Woolwich.

Resistance to Torsion and Shearing. 145

Table collected from various sources, showing the relative strength of metals to resist torsion, wrought iron being 1.—

TABLE XXXIV.

Wrought iron	1·00
Cast iron	·90
Cast steel	1·95
Gun-metal	·50
Brass	·46
Copper	·43
Tin	·14
Lead	·10

The above relative values must be considered as approximate only, because the resistance of different specimens of the same material varies more or less, according to its quality.

A few examples of the strength of crane shafts will be given, in the Chapter on Complex Structures. The nature of the stress to which such shafts are subjected is more like that which is produced in experiments, on the absolute strength to resist torsion, than is the case in mill work, where strength and stiffness become complicated with the question of motion and velocity. The strength of a shaft required to transmit power depends entirely upon the speed at which it is driven; a shaft that will convey 10 horse-power at a velocity of 50 revolutions per minute, will be strong enough for 30 horse-power at 150 revolutions, and so on in proportion.

Example.—To find the horse-power that can be transmitted by a good wrought-iron shaft, 3 inches diameter, driven by a wheel 2 feet in diameter, at a speed of 150 revolutions per minute. From the Table, we find that a 1-inch shaft will break with a stress of 800 lbs. acting at the end of a 12-inch lever, or at the pitch-line of a wheel 2 feet in diameter, and as the strength increases as the cube

L

of the diameter, a 3-inch shaft will break with 21,600 lbs. acting at the periphery of the driving wheel.

Let it be assumed that we are not to strain the shaft to above ⅙th of the breaking-weight, then we may have a constant load of $\frac{21600}{6} = 3,600$ lbs. acting upon the wheel.

The wheel, in making one revolution, travels over $2 \times 3\cdot1416 = 6\cdot28$ feet, and in running one minute $6\cdot28 \times 150 = 942$ feet.

Then the number of foot-pounds that can be transmitted by the shaft will be the load of 3,600 lbs. multiplied by the distance through which it travels—namely, 942 feet; thus, $3,600 \times 942 = 3,391,200$, and this number divided by 33,000 $= 102\cdot76$, which is the number of actual horses-power that can be transmitted by the shaft, under the given conditions of speed and factor of safety.

In the foregoing example, it is assumed that the load upon the driving-wheel is constant, which is true for ordinary shafts, but, in the case of an engine crank-shaft, such a condition cannot hold, because, even when the pressure on the piston is uniform, there are some points in the revolution at which the twisting force is practically equal to the full load upon the piston, while at other points the twisting force is nil, although of course the horse-power exerted by the piston and crank are equal.

The average pressure upon the crank is equal to the pressure on the piston, multiplied by the distance which it travels in making a double stroke, divided by the distance through which the crank-pin travels in the same time; now, the piston passes through a distance equal to twice the diameter of the circle described by the crank-pin, whereas the crank-pin travels over a distance equal to $3\cdot1416$ times that diameter, and hence the mean strain on the piston equals $\frac{3\cdot1416}{2} = 1\cdot57$ times the driving pressure upon the crank.

The varying strain thrown upon the crank-shaft is made, approximately, uniform by the fly-wheel, and hence second

Resistance to Torsion and Shearing. 147

motion shafts may be said to be uniformly loaded throughout, but it is evident that the portion of the crank-shaft, between the crank and the fly-wheel, has to withstand a greater strain than the part on the other side of the flywheel, in the ratio of 1·57 to 1, and it has to be made stronger to resist it.

In the case of engines working expansively, the horsepower is governed by the mean pressure upon the piston, but the crank-shaft has to be of sufficient strength to withstand the greatest pressure on the piston.

For example, if we had an engine working with the full pressure of steam during the whole of the stroke, and another of precisely equal power working with the same steam pressure at the commencement of the stroke, but cut off at ⅓ of the stroke, the mean pressure on the piston of the latter engine would be ·846 of that upon the other piston; and hence to enable them both to perform the same amount of work, the area of the piston of the latter would have to be 1·18 times greater than that of the former, and the maximum strain on the crank-pin and crank-shaft of the expansive engine would be 1·18 times that upon the crank-pin and shaft of the other engine, although the horse-power transmitted through the two shafts is precisely equal in each case.

It is necessary, therefore, in ascertaining the proper size for a crank-shaft, to find first the maximum pressure brought to bear upon the shaft, then multiply that pressure by the number of times the breaking strength is to exceed the working strength, and find the size of shaft, as explained for the ordinary shaft.

Example.—Required the diameter of the crank-shaft for a horizontal engine, which is to be worked with a pressure of 45 lbs. per square inch throughout the stroke, the diameter of the cylinder being 36 inches, and the stroke 5 feet, and the breaking-strength to be six times the working-load.

1. The maximum pressure on the crank-pin, exclusive of

friction, will be the pressure per square inch upon the piston multiplied by its area, thus, $45 \times 36^2 \times \cdot 7854 = 45,804$ lbs.

2. The breaking-strength $= 45,804 \times 6 = 274,824$ lbs. The leverage at which this maximum pressure acts is $\frac{9}{4} = 2\frac{1}{4}$ feet, and the weight required to break a 1-inch bar when acting at the end of a $2\frac{1}{4}$-feet lever $= \frac{8 \cdot 0}{9} = 320$ lbs.; but we have to provide for a weight of 274,824 lbs., and hence we must have $\frac{274824}{320} = 859$ times the resisting power of a 1-inch bar, and as the strength varies as the cube of the diameter, the cube-root of 859, which is equal to 9·5, will be the diameter in inches of shaft required.

To take another example. Let it be required to find the diameter of the crank-shaft, for an engine equal in power to that in the previous example, worked at the same pressure at the commencement of the stroke, but with the steam cut off at half stroke, the length of stroke of the piston being equal in both cases.

It has been shown that the area of the piston, and hence the maximum pressure upon the crank, would be 1·18 times greater than in the former engine; therefore we must provide for $274,824 \times 1 \cdot 18 = 324,292$ lbs. acting on the same leverage, and the diameter of the shaft required will be the cube-root of $\frac{324292}{320}$, which is equal very nearly to $10\frac{1}{4}$ inches.

These few examples will serve to show that it is necessary to know something more than the horse-power of an engine, before we can determine the proper size of the crank-shaft, for, although the horse-power is precisely the same in both engines, the one requires a shaft $\frac{3}{4}$ths of an inch larger in diameter than the other.

In each of the foregoing examples, the shafts have been assumed to possess sufficient torsional stiffness, and they would do so in the case of the crank-shafts which are short, and the 3-inch shaft might be able to transmit the stated number of horse-power, but it might twist through so many degrees that it would be quite unserviceable for driving machinery. There are likewise other points to be looked

Resistance to Torsion and Shearing.

at, beside the horse-power to be transmitted, even with ordinary shafting, namely, the distance from the driven end of the shaft to the point at which the power is taken off, for a shaft might be quite strong and stiff enough to drive certain machinery, provided the power was taken off near the driver, but the case would be widely different if the same power were taken off the shaft, say at 300 feet, instead of at a comparatively short distance from the driver.

The torsional stiffness of wrought-iron shafts varies as the fourth power of their diameter divided by the length, and it is found that wrought-iron shafts above $4\frac{1}{2}$ inches diameter, which are strong enough to resist the torsional strain, will be found sufficiently stiff to do their work; but below that diameter the stiffness of the shaft is the weak point, hence deflection must be otherwise provided for, in order to afford the necessary strength; therefore, a shaft of larger diameter than is absolutely necessary to resist the torsional strain must be used, in order that it may be sufficiently stiff for the steady driving of the machinery, so that in determining the proper size to be given to a long line of shafting, all these important points should be taken into account.

Resistance to Shearing and Punching.

The resistance of materials to the action of shearing and punching has not yet received so much attention as some of the other more important modes of resistance. Still, sufficient is known of the amount of force necessary to shear or punch wrought iron, to enable the student to form an approximate estimate from their relative tenacity, of the shearing and punching resistance of other metals, which have not yet been operated upon to the same extent.

The force which is required to shear or punch, depends on two conditions: first, upon the extent of surface which is to be shorn or broken through; and secondly, upon the nature or kind of material that has to be operated upon.

The operation of shearing or punching is not strictly a cutting, but a detruding action, and the whole work is done almost at the commencement or immediately after the first yielding. That some time is occupied in shearing or punching is largely due to the inherent elasticity of the material, the punch, and the machine, but when the elasticity and slackness are used up, then the resistance offered by the material is at once overcome, and the metal is detruded, broken, or pushed off, rather than cut through, and when the operation is performed, the resilience of the apparatus in assuming normal conditions is accomplished with a jerk.

From this it will naturally be inferred, that the process of shearing and punching requires a greater immediate mechanical force to accomplish the object, in that way, than would be required to cut the portion with an incisive instrument by a more gradual cutting or perforation. There is, therefore, from such an action, a constant repetition of sudden strains coming upon the tools, which, although lasting only for a second, yet when united in effect, tend to rupture the apparatus, unless it is made exceedingly strong, and this accounts for the ultimate breakage which generally overtakes all such machines, and which is fully explained by the results, shown in the Tables, relating to the elasticity of cast iron.

The resistance that is offered to shearing is a fraction under that of punching, and both are found to be a little less than the tensile strength of a piece of the material of equal area ; and from some experiments made on shearing and punching wrought iron, it appeared that, with fair uniform bearing instruments, to punch a hole one inch diameter in an inch plate, or to shear through a bar three inches broad by one inch in thickness, required nearly 80 tons of stress, the stress being as the metallic surface disturbed or broken through, which in these two cases would be nearly the same. With a half-inch plate or a bar half an inch thick, the stress would be about the half of 80 tons, and so on in proportion.

This amount of stress is considerably higher than that which appears to have been obtained, in some other published experiments, which go to show that the force required to shear or punch is about 80 per cent. of the tensile strain. But in making such experiments, however carefully, the result will much depend upon the precise condition of the acting surfaces of the punch or shears; for unless these surfaces have a perfectly flat and fair bearing, at the commencement of contact of the instruments, the work of the shearing or punching will be performed in detail, and thus the apparent strain will be reduced by being spread over a longer time. Hence arises the great practical advantage which is found to result from making the face of such punches so as not to have a fair bearing at the commencement, but slightly out of parallel with the bolster, and likewise the arrangement of the shears so as to act like a pair of common scissors, which give a gradual shear.

When it is said that the resistance to a shearing action is nearly the same as the resistance to tensile strain, it has to be borne in mind that the effect will depend on various conditions, for there are many forms of construction which may and do call the shearing action into activity, unintentionally, from some modification of the mechanical arrangements; to take, for example, the case of long links, in which the connection is formed by means of an eye or hole at the ends, into which a pin or bolt is inserted. It might be inferred that the bolt should have the same sectional area as the body of the link; but it is found that when so made, the round pin acts as a blunt wedge within the hole, and thus tends to tear it open. To give uniformity of strength, it is found necessary to provide a larger bearing surface, in order to counteract this detrusive tendency, and it is for this reason that some engineers have the bolts or pins of such structures, apparently out of proportion, generally to the extent of half a diameter, but this is true economy, notwithstanding.

The practical question is frequently raised, in regard to the comparative merits of punching and drilling, in perforating the rivet holes in boiler plates and for similar purposes. Punching is the cheaper process, and has the advantage of finding out and breaking an exceptionally bad or brittle plate, but, on the whole, it is not so efficient as drilling. The drill, being a cutting instrument, deals more gently with the material, and the plate, so treated, is more likely to maintain its full strength after the operation. As the line of rivets is the weakest part of the structure, it seems throwing away an important advantage not to drill the holes, even if it is a little more expensive to do so. It is probable, however, that if the drilling system were established, by proper organisation of plant and well-directed arrangements, it would not be a more costly process than punching. Already some of the larger engineering houses are drilling their boiler plates, and are able to make boilers almost as economically, on that system, as by the older plan of punching.

From some valuable experiments made in 1858, which are recorded in the Proceedings of the Institute of Mechanical Engineers, and were published in the 'Artizan' of that year, it was found that the stress required to punch a hole one inch in diameter, through a wrought-iron plate $\frac{1}{2}$ an inch in thickness, was equal to 36 tons, and to force the same punch through a plate of double the thickness required 69 tons, which is not quite the double of the former result. The mean of the two experiments gives 22·5 tons, per square inch of detruded sectional area.

A punch 2 inches in diameter was pushed through three plates in succession, the respective thicknesses being $\frac{1}{2}$ an inch, 1 inch, and $1\frac{1}{2}$ inch, and the varying pressures required were 65, 132, and 186 tons. This gives a lower mean than the former—namely, 19·4 tons per square inch of detruded

Resistance to Torsion and Shearing. 153

sectional area—thus showing that, by increasing the size, the proportionate stress diminishes, as in the above examples, to the extent of 14 per cent.

The following Table contains the results of another set of experiments, made with actual weight on the end of a lever, the weight being gradually applied, until the iron was detruded :—

TABLE XXXV.

Diameter of punch. Inch.	Thickness of wrought-iron plate. Inch.	Sectional area of detrusion. Square inches.	Total weight bearing on punch. Tons.	Stress per square inch of area. Tons.
0·250	0·437	0·344	8·384	24·4
0·500	0·625	0·982	26·678	27·2
0·750	0·625	1·472	34·768	23·6
0·875	0·875	2·405	55·000	23·1
1·000	1·000	3·1416	77·170	24·6

In the several experiments, the average pressure required for punching seems to be greater for thin plates than for thick plates, which is probably due to the greater relative proportion of hard exterior surface in the thinner plates.

From a comparison of experiments, in punching and shearing, the stress required to detrude, per unit of sectional area, appears to be nearly the same in both cases, with moderately small holes; but with large holes there is a perceptible difference, nearly 5 per cent. less pressure being required to shear than to punch a given sectional area, which is probably owing to the effect of friction at the entrance of the punch.

For comparison, the following Table, taken from the Proceedings of the Institute of Mechanical Engineers, will show the stress required for shearing :—

TABLE XXXVI.

Direction of shear.	Sectional area of surface detruded.		Pressure on the detruding instrument.		Remarks.
	Thickness and breadth in inches.	Area in square inches.	Total pressure in tons.	Stress per square inch of area detruded in tons.	
flat	0·50 × 3·00	1·50	33·4	22·3	Mean stress.
edge	0·50 × 3·00	1·50	34·6	23·1	} 22·7
flat	1·00 × 3·00	3·00	69·2	23·1	
edge	1·00 × 3·00	3·00	68·1	22·7	
flat	1·00 × 3·02	3·02	59·7	19·8	} 21·5
edge	1·00 × 3·02	3·02	62·1	20·6	
edge	1·80 × 5·00	10·20	210·6	20·6	

The foregoing experiments were made with the shearing instruments parallel, when the work of detrusion was done after the first pinch. It was found that by spreading the operation over a longer period by means of inclined instruments, the stress required was considerably reduced, as will be seen by the following Table :—

TABLE XXXVII.

Size of bar in inches.	Stress with bar laid flat upon its side, in tons per square inch.	Stress with bar laid on edge, in tons per square inch.	Percentage of less stress required when detruded on flat surface.
3 × 1½	18·2	20·1	10
4½ × 1⅝	14·3	17·9	20
3 × 1	15·7	21·1	26
5¼ × 1¾	16·7	22·6	26
6 × 1½	15·0	18·4	18

The above experiments were made with shears, the blades of which were at an inclination of 1 in 8. The less stress required when the same bar of iron is cut on the side, is

Uniformity of Sectional Area. 155

most probably due to the circumstance that the work to be done is spread over a longer period.

By comparing the stress per square inch in Table XXXVII. with that in Table XXXVI., we obtain the averages given in the following summary of results : —

Description of shearing blades.	Mean force required to shear bars on flat, in tons per square inch.	Mean force required to shear bars on edge, in tons per square inch.
Parallel	21·7	21·7
Inclined	16	20

Thus teaching four important lessons. First, that with parallel shears, the shearing force required is equal in either position of the bar. Second, that with inclined shears, the shearing force is 20 per cent. less for the bars on the flat, than it is for the same bars on edge. Third, that when the bars are cut when placed on edge, only 8 per cent. is saved by using inclined shears. Fourth, that when bars are sheared on the flat, a saving of 26 per cent. is effected by the inclined, as compared with the parallel shears.

CHAPTER XI.

ON THE IMPORTANCE OF UNIFORMITY OF SECTIONAL AREA.

IN designing machines or structures, not only is there a correct form, but it is of the greatest importance, that that form should not be departed from unnecessarily. The correct form is that, in which every part is proportioned to the straining action to which it is subjected, so that the stress reaches the same maximum limit on every section. If, instead of being thus proportioned, a structure has a surplus of material in some parts, that surplus may not only not strengthen it, but may positively weaken it, and may render parts, otherwise of sufficient strength, incapable of sustaining the load for which they have been calculated. Thus, for example, a chain with one link smaller than the others, a shaft with one journal of insufficient diameter, or a tie-rod nicked at any point, is reduced in strength, not only in proportion to the reduction of section at the weak point, but in a still greater degree. In consequence of the surplus resistance in the other parts, most of the elastic work will be concentrated on the weak link, the weak journal, or the nicked section of the tie, as the case may be, and those parts will gradually be more weakened, and at last overcome, before the other parts of the structure are affected.

Even before the principle just indicated was fully understood, it was more or less anticipated by experienced engineers; and it is interesting to recall the intuitive skill in construction, often displayed in the forms adopted for the details of structures. For example, in parts of early locomotive engines, where cotter holes or keyways were formed, an increased depth or thickness was generally given at that point, so as to make up for the material cut away. By this means the whole structure was brought approximately to equality of resistance, and the concentration of the elastic work on a particular section was prevented.

Although, strictly speaking, we cannot increase the strength of a structure by securing equality of sectional area, because the strength still remains dependent on the weakest section, yet virtually we increase the strength, by preventing the gradual deterioration of the resistance at one point, and thus diminishing the risk of fracture.

As a simple example, let us consider an ordinary screw bolt. A screw bolt may be formed, either with the screw thread in relief, the shank being of the same diameter as the screwed part at the bottom of the threads, or the screw may be cut into a bar, and the shank is then of the diameter of

Uniformity of Sectional Area. 157

the screwed part, measured to the outside of the threads. Two bolts may be thus formed, of the same section at the weakest part, but the former bolt will yield equally throughout when loaded, while the latter will yield chiefly at the point of junction of the screw thread with the shank. Hence, when subjected to stress, the stretching will be distributed in the former, but in the latter the screw threads will be drawn out of pitch, and will no longer match the nut. If these two bolts are treated alike, say by using a screw key with a long handle, so as to twist off the nut by tension, or by combined tension and torsion, then the former bolt, although lighter in weight, will yet require more turns of the screw key before it will give way than the latter. Practically it is the stronger, and it is important that this should be understood, because the same principle is applicable in many other cases.

It is not essential that the screw thread should be in relief, with a small shank, the uniformity of sectional area may be maintained in other ways. For many purposes it is convenient that the shank of the bolt should fit the bolt hole, and the bolt hole must be large enough to admit the screwed part. Hence, in these cases, the shank of the bolt requires to be maintained equal in diameter to the outside diameter of the screw thread. To secure uniform section in the bolt, various plans have been adopted. In some bolts, the shank is made hollow from the head up to the neighbourhood of the screw thread. These bolts have been used in fixing armour plates. In other bolts, the shank has been made of cruciform section, or of a square or triangular section with rounded corners. All these plans give satisfactory results, so far as depends on the uniformity of the section.

The condition of uniformity of section, in practical operations, is liable to be modified, either for the better or the worse, in many ways. For example, the effect produced in the forming of the screw thread of bolts such as those before referred to, may be seriously injured unintentionally, from the mode of making, cutting, or forming the thread, as depending on the sharpness or bluntness of the screwing dies. With good instruments, which do really cut out the thread, and do not distress the internal structure of the material, the bolts are more reliable than those that are formed with blunt dies, which act more as abrasive, or even detrusive instruments, than as genuine cutters. The effect of such violence of treatment is to harden or loosen the iron, and to render the bolt more brittle. With good instruments and ordinary care in the manufacture, and when the form of bolt section has been arranged on correct principles, the strength of bolts is about equal to that of the iron out of which they are made, ranging between 20 and 25 tons per square inch of sectional area at the weakest point, according to quality.

Similar remarks apply to the strength of chains. The disastrous effect of a weak link is seen, more especially, in those chains that are used for cranes and in slings for lifting heavy weights. As a rule, such chains are proved up to about the half of the ultimate tensile strength, or about the double of the intended ultimate working load, nevertheless it is frequently found that, after working for a short time, some weak link begins to stretch, and if this is not observed in time, it will break unexpectedly. Such a link must have been weak originally, and, during its working career, it must have had more than its fair share of the elastic duty to perform; hence if it is not cut out of the chain in time, it will ultimately break with a considerably less load than that which it was intended to support, or less than the quarter of its original ultimate strength in its early days; on the other hand, when a chain is so well made that all its links are of uniform strength, or nearly so, the endurance is much greater, and it will go on for an indefinite period without breaking. At the same time, a change for the worse overtakes all chains in course of time, that is, if they are in constant use; and, in consequence, it is a rule in the War Department, that all

Uniformity of Sectional Area.

chains of cranes or slings are to be drawn through the fire, and thereby annealed, periodically, which has the effect of restoring the quality, by putting the material of the links into a condition of equilibrium. By this course, their life is protracted indefinitely. After the chains are annealed, they are then subjected to a proof strain, which is the half of the ultimate strength, and double the usual working load, according to size, which is as under:—

TABLE XXXVIII.
Table of Proof Strains.

Size of iron in chain, in inches.	Proof strain, in tons.	Size of iron in chain, in inches.	Proof strain, in tons.
$\frac{5}{16}$	$1\frac{1}{8}$	$\frac{13}{16}$	$7\frac{7}{8}$
$\frac{3}{8}$	$1\frac{5}{8}$	$\frac{7}{8}$	$9\frac{1}{8}$
$\frac{7}{16}$	$2\frac{1}{4}$	$\frac{15}{16}$	$10\frac{1}{2}$
$\frac{1}{2}$	3	1	12
$\frac{9}{16}$	$3\frac{3}{4}$	$1\frac{1}{8}$	$15\frac{1}{4}$
$\frac{5}{8}$	$4\frac{5}{8}$	$1\frac{1}{4}$	$18\frac{3}{4}$
$\frac{11}{16}$	$5\frac{5}{8}$	$1\frac{3}{8}$	$22\frac{5}{8}$
$\frac{3}{4}$	$6\frac{3}{4}$	$1\frac{1}{2}$	27

Chains are of two sorts: first, the open link chain, which is commonly used for cranes and like purposes; and second, the stayed link chain, which is used for cables and some other purposes. It is to be noted that, whatever the explanation may be, the stayed link, when made of the same iron as the open link, is stronger than the other, nearly in the proportion of 9 to 6. To take, for example, an inch chain, that is to say, a chain made with iron one inch in diameter, the ultimate strength of the open link is about 24 tons, while that of the stayed link may reach 30 tons. The office of the stay is to prevent the collapse of the link, and thereby to intercept the shearing action due to the wedge action of the one link within the other. The great practical objection to the stayed link chain is its weight, and its extreme roughness in working over pulleys.

A good approximate rule, for mentally estimating the strength of chains, is to remember the strength of one particular size—say of one-inch chain—the safe working load of which is 6 tons for the open link, and 9 tons for the stayed link: the comparative strength of other chains is to that of the inch chain, as the squares of the diameters of the iron out of which they are severally made. This does not hold strictly correct with very large, nor with extremely small sizes, which both lose strength to a small extent, the former from the iron not being quite so good, and the latter from the welding not being quite so perfect.

Another simple rule, for the approximate safe working load of chains of different sizes, is to square the number of eighths of an inch, in the diameter of the iron out of which the link is made, and to strike off the last figure as a decimal. For example, in the inch-chain there are 8 eighths of an inch.

$$8 \times 8 = 64,$$

or, striking off the last figure, 6·4 tons, the greatest safe load.

Again, for a $\frac{3}{8}$ inch chain, $3 \times 3 = 9$. Striking off the last figure we get 0·9 tons for the greatest load consistent with safety.

Uniform and proportional sectional area adds to the strength of beams, when subjected to transverse strains. Any notch made on the under side of a beam for the insertion of bolts, nuts, or washers, should be studiously avoided.

In the construction of heavy timber roofs, where diagonal bolts or tie-rods have to be introduced, at an angle, through the principal beam, some persons are so unskilful as to make a notch on the under side of the beam, for the nut and washer. Such an arrangement is most objectionable, because the damage done to the strength is permanent. The proper course, under such circumstances, is to employ a broad iron washer-plate, with its flat surface bearing upon the under side of the beam, the washer-plate having a hole cored out, at the proper angle for the reception of the bolt, and with a corresponding boss, having the bearing surface for a nut and supplementary washer, at right angles to the axis of the tie-rod.

Another objectionable practice sometimes seen, in connection with rough structures, is the formation of slits or cotter-ways in bolts, and in having that part of the bolt made with the same exterior diameter as the solid portion of the main body. Such a bolt would be really stronger, for all practical purposes, if the main body was reduced to a smaller diameter, equal in sectional area to the part in which the slit has been made. The proper course is, to upset the bolt at the part which has to be punched, and then to form the cotter-way by taper punching, in order to leave the metal entire, and thus obtain uniformity of sectional area.

The Strength of Ropes.

The strength of hempen ropes is found to depend primarily on the quality of the hemp fibre with which they are made. The fibres vary from 3 to $3\frac{1}{2}$ feet in length, and a number of them are twisted together to form a yarn, this twisting introduces the element of friction, which effectually prevents the fibre from being drawn asunder. When a rope is twisted it necessarily becomes shorter than the strands of which it is made, and the amount of twist given to ropes is usually expressed by the extent to which they are shortened; thus if shortened one quarter of their length, the twist is then said to be one quarter. The nominal size of ropes refers to the circumference, thus a six-inch rope is a rope of six inches circumference.

As a rule, new white ropes are stronger and more pliable than ropes made with tar. The tarred rope, however, retains its original strength for a longer period, especially when exposed to wet. Tarred ropes spun hot are stronger than tarred ropes spun cold, and are more impervious to water.

The ultimate strength of ropes is usually considered to be about 6400 lbs. per square inch of sectional area. From some recent experiments made at Woolwich, the strength was found to range from 9874 lbs. in a 9-inch to 10,783 lbs. in a 2-inch rope. The same series of experiments showed that Italian hemp ropes are stronger than those of Russian hemp.

The Woolwich experiments teach also three important lessons to the student:—1st. That there is much variation in the strength of pieces of rope even when cut from the same coil. Thus, the 4-inch ropes range in tenacity from $5\tfrac{5}{8}$ to $7\tfrac{15}{17}$ tons. The 6-inch ropes from $14\tfrac{1}{4}$ to 17 tons. The 9-inch ropes from 25 to $29\tfrac{11}{18}$ tons. The minimum strength observed in the experiments is that which should be relied on in practical calculations. 2nd. That there is considerable loss of strength from the tear, wear, and exposure to the weather during a few months working. Thus the strength of an apparently good 6-inch Italian hemp rope, after working, was only $10\tfrac{3}{4}$ tons, as compared with $14\tfrac{1}{4}$ tons, the strength of a new rope. The old 6-inch Russian hemp rope broke with $5\tfrac{1}{2}$, while the new one required $11\tfrac{1}{4}$ tons. This fact suggests that we ought to allow a large margin between the working safe load and the ultimate strength of new ropes. Thus, we may make the safe load $\tfrac{1}{8}$th of the ultimate strength. 3rd. That a double rope is in certain cases weaker than would be expected from experiments with single ropes. All the double rope slings, suspended from an ordinary crane hook, broke with less than double the strength of a single rope. On the other hand, when thimbles or pulleys were used, then the double ropes nearly always required double the maximum load which would break a single rope.

PART II.

ON THE STRENGTH OF STRUCTURES.

CHAPTER XII.

BEAMS AND GIRDERS.

HAVING described the physical properties of the various materials employed by the engineer, we have next to consider those structures, either simple or complex, the strength of which depends on the form, arrangement and construction, of the parts, as well as on the material of which they are composed.

In studying the laws of the resistance of structures, a knowledge of elementary mathematics is necessary. In the following chapters, the subject is treated in the most simple manner, in the hope that they may thus be rendered useful to all classes of readers, and especially to those whose mathematical knowledge is limited, by enabling them to calculate the strains in the more simple structures required in practical operations.

Every practical engineer should be able to ascertain, approximately, the strains which are likely to be brought upon the component parts of any structure, both by the weight of the structure itself, and the load which it is intended to support. He should likewise know the amount of stress which may, with perfect safety, be put upon each square inch of section of the material of which the structure is composed. It will then become a matter of simple proportion, to ascertain the number of square inches of any given material, that should be introduced into each part of the structure, in order to enable it to resist with safety the strain which it will be required to sustain.

Of late years, unfortunately, too much reliance has been frequently placed upon the formulæ which are to be found in different books; these are drawn from a variety of sources, and are generally given without the data from which they were originally deduced. This reliance upon mere formulæ is a great mistake, and often leads to expensive errors. In some cases, formulæ, originally intended for one kind of structure, have been applied in designing a different kind of structure, and the incorrect result arrived at has only been discovered, when too late to avoid the consequences of the mistake.

In the following chapters, the natural laws of mechanics are brought into prominence, so that the student may be led to use his judgment in the application of rules. At the same time, a general knowledge of mathematics will be found of the greatest value, giving a more comprehensive grasp of the whole subject, and fitting the student to use formulæ with intelligence.

From a very early period, it was known that the rectangular beam, placed on its edge, would carry safely a heavier vertical load than the same beam placed upon its side. In the same way, it was understood that a given quantity of material, arranged in a hollow form, was stronger than when in the solid shape. The reasons why these things are so, are now also well understood, and the practical engineer should make himself acquainted with them, if he would carry out works with safety and economy.

The present chapter will relate chiefly to beams and girders, and the principles applicable to these are mostly derived from the investigations and experiments of Hodgkinson, Fairbairn, and others. An effort has been made to explain these principles in as simple a manner as possible; but the full treatment of the subject would involve a higher degree of mathematical knowledge than is here assumed. The student who wishes to pursue the subject further is recommended to consult the treatise by Professor Twisden, in the same series as the present volume.

Beams or Girders.

The term 'beam' is generally applied to any large piece of material, which carries a load at a distance from the point or points of support upon which it rests, and which is thereby subjected to transverse strain. But this term, conjoined to another word, is often used for certain parts of structures, which in reality are subjected to longitudinal strains only, such as, for instance, 'tie-beam' or 'straining-beam,' both of which are subjected to tensile strains only.

The term 'girder' is now almost universally adopted by engineers, as the name for beams, which are supported at both ends and subjected to transverse strain, and the term 'cantilever' is generally used for beams, which are supported at one end only and subjected to transverse strain.

Girders are made of many forms, depending on the material employed. Those in most common use are of the rectangular, flanged, or tubular forms of section; the flanged girder is indiscriminately made of cast, or wrought iron, and the term includes all forms in which a flange or flanges are connected to a single vertical web, whether the web is a solid plate or an open lattice, or whether the flanges are parallel or otherwise.

In tubular girders, which are mostly built up of wrought iron, the two flanges are connected by two webs, and in the case of the celebrated tubular bridges of Fairbairn, the tubular girders are of such large dimensions, as to admit of the load being carried inside the tube, instead of on the outside in the usual manner.

In considering the strength of beams or girders, when subjected to transverse or cross strains, three chief points present themselves for consideration, namely :—

First, the mechanical effect which any given load produces upon the section of fracture, of whatever form, under varying conditions of support.

Second, the nature of the strains that are brought to bear upon the girder, and the manner in which its resistance to those strains is affected by the form of the section of the girder.

Third, the actual strength of the material composing the girder, which is to be thus strained, and which can only be determined by direct experiment, or by placing reliance on the numerous published experiments, which have been so carefully carried out, by Barlow, Fairbairn, Hodgkinson, Kircaldy, and many others.

The relative strength of a girder, so far as it depends on the manner in which it is loaded and supported, can be expressed very simply. Taking a given girder, of a given span, the load may vary in the ratio of 1 to 8, according to the different ways in which the girder is supported and the load distributed.

Position of Support and Load.

	Relative strength.
When supported at one end and loaded at the other	1
When supported at one end and load distributed	2
When supported at both ends and loaded at the centre	4
And when supported at both ends and the load equally distributed	8

When a uniform beam is securely fixed at both ends, by encastrement or otherwise, instead of merely resting on supports, it is then under an entirely different set of conditions. Theoretically it would break, simultaneously, at three different sections, instead of giving way in the middle only, but practically this would never occur, unless the fixing of the ends were exceedingly firm, as, for instance, when the beam and the abutments were part of the same casting. The increase of strength, due to secure fixing of the ends, has been variously stated; according to some, the relative strength of

Beams and Girders.

the beam above referred to would become equal to 12, and, according to others, even to 16; but it is obvious that the value will depend on so many conditions, that it is impossible to say precisely what it is, beforehand; it must, therefore, be considered as doubtful, unless all the conditions are known definitely: at the same time the importance of secure fixing, wherever it can be employed should be noted.

The reason why the same girder possesses the above relative strengths, under the given varying conditions, may be very clearly shown, by supposing the girder in question to act as a bent lever, one arm of the said lever being equal in length to the distance from the point of support to the weight, and the other arm of the lever to be the depth of the girder.

In considering the strength of such a girder, in which the amount of deflection due to the load is inconsiderable, we may neglect the small curvature of the beam when loaded, and consider that the straining forces act perpendicularly to the longitudinal axis of the beam. They are at right angles to the fibres of which it is composed, if it is a timber beam, and at right angles to the fibres, of which we may imagine it to consist, for theoretical purposes, if it is a metal beam. But the longitudinal tensions and compressions in the beam act parallel to the axis, or in the direction of the fibres of which it consists or may be imagined to consist. The loading forces act at right angles, therefore, to these tensions and thrusts.

Let figs. 16, 17, 18, 19 represent four similar flanged-girders supported and loaded under the above-stated conditions. We know, by experience, that a weight acting at the end of a cantilever, as shown in fig. 16, will tend to bend the free end of the girder in a downward direction, as shown by the dotted lines a, b, c, which represent a bent lever; we also know that, in order that such a lever may be in equilibrium, the product of the weight at the end of one arm, multiplied by the length of the arm, must be equal to the product of the length of the other arm, multiplied by the weight acting upon it.

168   *On the Strength of Structures.*

Supposing, as an example, that a weight of 1 ton is required at the free end of the girder to break it, and that the

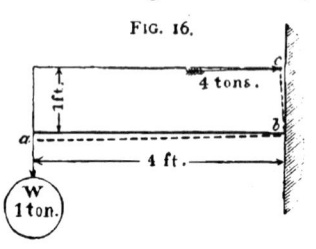

FIG. 16.

[Figure 16: cantilever girder 4 ft long, 1 ft deep, with 4 tons at support end and W = 1 ton at free end]

girder is 4 feet long and 1 foot deep, then the weight or force, required at the other end of the lever, will be the product of the weight of 1 ton, multiplied by 4 feet, the length of the long arm of the lever, and divided by 1 foot the length of the other arm, and will be equal to 4 tons.

The same girder (fig. 17), supported in a similar manner, but with a distributed load, will break with a weight of two

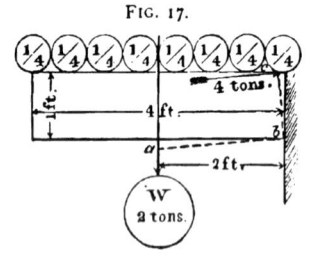

FIG. 17.

[Figure 17: same girder with distributed load of eight ¼-ton weights, 4 tons at support, W = 2 tons acting at 2 ft]

tons, or twice the weight required to break the first girder, because the mean distance at which the weight acts upon the beam, or, in other words, the centre of gravity of the weight will be only $\frac{1}{2}$ the distance from the support that it is in the former example, and hence it will have a leverage of 2 feet instead of 4 feet.

The breaking strain, or the force required at the end of

Beams and Girders. 169

the short arm of the lever, is of course the same as in the first example—namely, 4 tons—and the weight at the end of the long arm, necessary to produce the strain of 4 tons, will be 4 multiplied by 1 and divided by 2, that is 2 tons, or twice the weight required to produce the same strain when placed at the end of the cantilever.

A similar girder (fig. 18), supported at both ends and loaded at the centre, will require 4 tons to break it, or 4

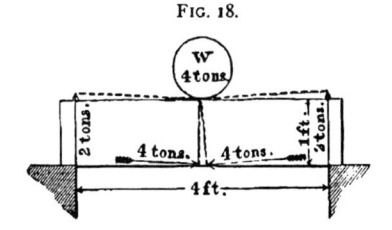

FIG. 18.

times the load required to break the first beam. The weight upon a girder, supported at both ends, is transmitted with undiminished force to its supports, and each support will react upon the girder with a force equal to a certain portion of the total load, according to the position of the weight relatively to the supports.

The amount of each reaction is determined by the principle of the lever; the reaction of either support bears exactly the same proportion to the total weight, that the distance from the weight to the opposite support bears to the total distance between the supports; for example, if we suppose a weight of 4 tons to act upon this girder, at a distance of 1 foot from one support, the distance to the opposite support will be $\frac{3}{4}$ of the whole distance between the supports, and the weight upon the near support will be $\frac{3}{4}$ of the total weight, or 3 tons.

When the weight is at the centre of the girder, as shown in fig. 18, the distance from the weight to either support is $\frac{1}{2}$

170   *On the Strength of Structures.*

the total distance between the supports, and hence the upward reaction of either support is equal to ½ the total weight upon the girder.

The leverage of each of these reactions is equal to ½ the span of the girder, or 2 feet, and the force required to break the girder, when acting at the short end of the short arm of the same, is 4 tons as before, and therefore the reaction of each support must be 4 multiplied by 1 and divided by 2, which equals 2 tons; and the total weight required upon the girder to produce the reaction of 2 tons upon each support is 2 × 2 = 4 tons, and hence, the girder will not break, until it is loaded with 4 times the weight which would break the same girder, when supported and loaded as shown in fig. 16.

The same girder, loaded with an uniformly distributed weight, and supported at each end, as in fig. 19, would break

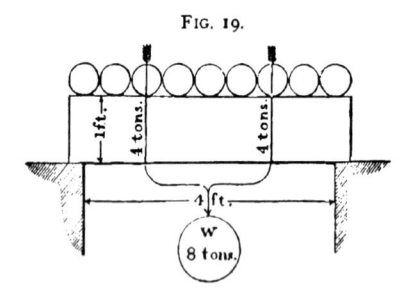

FIG. 19.

with eight times the weight required to break it, if supported and loaded as in fig. 16, or twice the weight which would break it if supported and loaded as in fig. 18.

With an uniformly distributed load, each particle of the load acts with a gradually increasing leverage from the support, where it is nil, to the centre, at which point it is equal to ½ the length of the girder, and the mean leverage of all the particles composing the weight upon each half of the girder is at a point midway between the centre of the girder and the support. Its effect upon the material at the centre of the girder is precisely the same, as if half the weight upon each half of the girder was placed at a point twice the distance from the support, which would be at the centre of the girder; or, to state the result shortly, an uniformly distributed load has the same effect at the centre of the girder as half the same load acting at the centre. An uniformly loaded girder will not break until it is loaded with twice the weight which would break it, when placed at the centre of its length.

The above has reference only to the theoretical conditions of beams, and does not take into account the weight of the beam. In the application and use of beams, practically, the weight of the beam itself must be considered. In small beams, of short span, the weight of the beam itself is small compared with the load it will carry. But in large beams, of long span, the weight of the beam itself becomes a very important element in the calculation. In long cast-iron beams, the weight of the beam itself may be equal to half the load it will carry, exclusive of its own weight.

The strength of a beam is likewise affected, by the manner in which the materials of which it consists are disposed in its construction. Many theories have been propounded by eminent mathematicians, but scarcely any of these theories fully explains the law which gives the results found by experiment. The knowledge derived from experiment is, as a rule, quite sufficient for the greater number of constructions in ordinary practice, and on this the following observations depend.

We have already seen that the girder tends to bend under its load, and hence the top layer of fibres must necessarily be compressed, and the bottom layer extended, and this compression and extension is not confined to the outer layers of fibres, but affects those in the interior of the girder to a gradually decreasing extent, as their distance from the top and bottom of the girder increases, until, when a certain point is reached, the fibres are neither compressed nor extended, and hence retain their original length.

This plane or surface of unaltered length, at or near the centre of the depth of the beam (when the material is nearly equal in tensile and compressive strength) is called the 'neutral surface,' and the line in which this surface cuts any transverse section of the girder is called the 'neutral axis of the section.'

The neutral axis of any section of a girder is, therefore, the line of demarcation between the forces of tension and compression exerted by the fibres crossing that section. In addition to these strains of tension and compression, there is a shearing strain, caused by the tendency of the weight to separate the part of the beam upon which it immediately rests from the adjoining part; this tendency is resisted by the material, and the weight is thus transmitted to the adjoining part, and by it to the next adjoining part, and so on to the supports. In parallel flanged girders, this strain is chiefly resisted by the web.

That there is no tensile strain upon the upper half of such a girder may be shown by an experiment. If we take a girder of the class here referred to, either of wood or metal, and cut it across its upper surface, and then insert a thin piece of wood or metal into the saw-cut, simply to prevent the opening from closing again when the weight is put on, the beam will be found to be as strong as before; but, on the other hand, the slightest notch on the under surface will weaken it considerably, and this will become apparent if the beam is tested.

It has been previously stated, that the strain upon the fibres of the beam gradually decreases from the top and bottom to the neutral surface of the beam, and hence the nearer the fibres are to that point the less they contribute towards supporting the load, and the farther they are removed from it, the greater is the power which they have to assist in its support; therefore, the neutral line is the part

Beams and Girders.

where holes, for fixing other parts connected with the girder, may be made with impunity.

In correctly proportioned flanged girders, the position of the neutral surface is considered to pass through the centres of gravity of the transverse cross sections, and the intensity of stress on each flange is directly proportional to the distance of the flange from the neutral axis. The section of each flange should be so proportioned, that the intensity of the stress exactly corresponds with the resistance of the material to tension or compression. It must, however, be borne in mind that the material of the compression flange is subject to distortion, and the real resisting power of this flange is its power to resist distortion or crippling, and not absolute compression.

The position of the neutral surface of a rectangular beam depends entirely upon the nature of the material; if its tenacity and power of resisting compression are equal, then the neutral surface will be in the middle of the depth, but not otherwise. Cast iron, for example, has about six times the power of resisting compression that it has of resisting extension; therefore the depth of material above the neutral surface will be to that beneath it as $\sqrt{1}$ is to $\sqrt{6}$, or as 1 is to 2·449, or, in round numbers, as 2 is to 5; and it would only bear the saw-cut $\frac{2}{7}$ of its depth, without loss of strength.

Let fig. 20 represent an exaggerated view of a cantilever loaded at its outer end, and $a\ b\ c\ d$ a portion of the beam before deflection; let the upper edge, after deflection, be extended from the length $b\ a$ to the length $b\ e$, and the lower edge compressed from $d\ c$ to $d\ f$, then the lines across the two triangles represent the alteration of length of the intermediate fibres, the neutral surface $g\ h$ dividing the depth of the beam into two equal parts.

The total stress upon the fibres composing each half of the beam is the product of the area of half the beam multiplied by the mean strain upon the fibres, and if the strain upon the extreme fibres is 4 tons, the mean strain upon all the fibres will be one-half 4 tons = 2 tons, and the total strain upon either half of the beam will equal 2 tons multiplied by the breadth of the beam, and by half its depth.

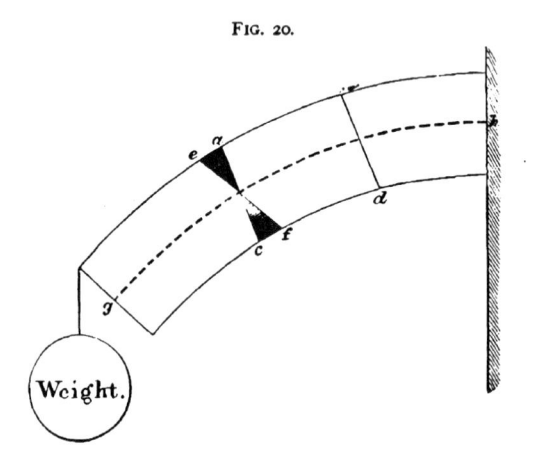

FIG. 20.

Further, as the lines across the triangles are proportional to the strains upon the several fibres, the centre of the tensions and compressions (that is to say, their resultants) will coincide with the centre of gravity of the two shaded triangles, which will be ⅔ of their height from the neutral axis, and hence the distance between the two centres of strain will be ⅔ of the depth of the girder; and, supposing one of them to act as a fulcrum, then the moments of the forces acting about that point will be, on one side, the product of the weight multiplied by its distance from the section of the girder in question, and, on the other side, the product of the tensile or compressive strain multiplied by ⅔ of the depth of the beam.

If we knew in all cases the position of the neutral axis

and the tensile strength of the material, we could then compute the strength of rectangular beams without any additional data; but without such precise knowledge, the best, or at least the easiest, method of determining the resistance of rectangular beams is to make a series of comparative experiments upon model beams, and determine their strength and deflection. From these experimental results a constant number may be deduced for each description of material. Such experiments have been made, and the table at page 182 contains the constant numbers for some of the materials which are most commonly used in works of construction.

The strength of flanged girders, of similar section, varies inversely as their lengths, and directly as their depths, and as the sectional area of their flanges; for if we double the length of either of the beams in the foregoing examples, then only one-half the weight upon the long arm of the lever would be required to break the beam; or if, on the other hand, we double the depth, we thereby double the strength, because the short arm is twice the length, and hence the flange, from its position, would be able to resist twice the weight acting at the end of the long arm of the lever; or if we double or increase the area of either flange, we also double or increase the strength in the same proportion.

It may be stated, as a general rule, that the strength of solid rectangular beams varies inversely as the length, directly as the breadth, and directly as the square of the depth, but up to a certain extent only—namely, so long as the said beam can be kept from twisting. For if we double the length of the beam, the resisting power of the section at any particular part will still remain the same, so that the actual weight required to overcome it will thus be reduced to one-half, because it acts now with twice the leverage; or, on the other hand, if we double the breadth of the beam, we thereby double the number of the resisting fibres, but do not alter the leverage, and consequently it has twice the strength; but if we double the depth, we then have twice the number of fibres and also twice the leverage, and therefore we have four times the resisting power. Hence, the strength is as the square of the depth.

The strength of square beams varies inversely as the length, and directly as the cube of the side of the square, because, if we double the side of a square beam, we then have four times the number of fibres or molecules, acting at twice the leverage, which gives a resistance equal to eight times that of the original beam; in other words, the strength is as the cube of their sides.

The strength of cylindrical beams varies inversely as their length, and directly as the cube of their diameter, for the same reason as that stated for the square beam.

The strongest solid rectangular beam, which can be cut out of a round log of timber, has a cross section, the square of the breadth of which is equal to one-third of the square of its diagonal. In other words, the proportion of breadth to depth is as 5 to 7 nearly. Hence, we have the following construction for cutting the strongest beam out of a round log.

FIG. 21.

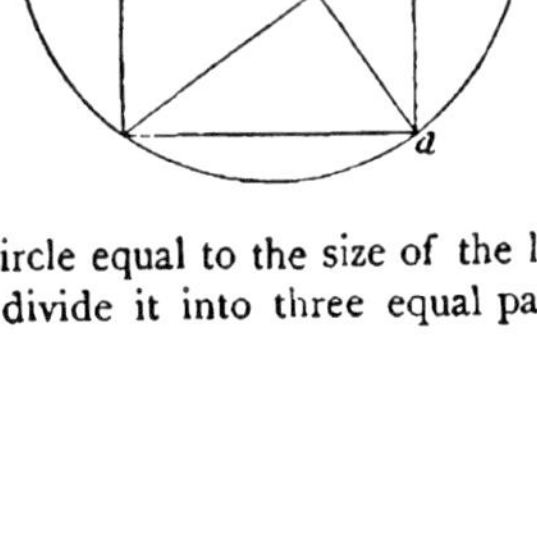

Describe a circle equal to the size of the log, and draw a diameter $a\,d$; divide it into three equal parts $a\,b$, $b\,c$, and *c d*; erect perpendiculars at *b* and *c* upon opposite sides of the diameter intersecting the circle, and then join the points in which the diameter and perpendiculars intersect the circle, and the rectangle so formed is the section of the strongest beam that can be cut from the log.

Rectangular beams are often made much deeper, in proportion to the breadth, than the beam given above, but they have to be kept from twisting by resorting to cross bracing or otherwise; still, for independent beams, the above proportion should always be employed where practicable. In some instances, the strength of the web of a cast-iron girder, considered as an independent rectangular beam, is added to the strength of the flanges, and a much greater apparent strength is thereby assumed than is justified either by experiment or in the practical constructions of the workshop.

The strength of a square tube when used as a girder, with its sides placed in a vertical position, is to the strength of a round tube of equal thickness and span, and having a diameter equal to the side of the square tube, as 17 is to 10. The strength of a similar round tube, having a diameter equal to the diagonal of the square tube, is to that of the square tube as 100 is to 85.

With square and round tubes of equal thickness and weight, their peripheries will be equal, and, when the sides of the square tube are vertical, their relative strengths are as 105 is to 100. The square tube, when placed with its sides in a vertical position, has thus a slight advantage over the round tube; but in such instances as crane-posts, which are subjected to transverse strains in all directions, by the load being slewed upon them all round the circle, the round form is preferable, because it is much stronger than a square tube of equal weight, when the load acts upon it in the direction of its diagonals, there being then so much more of the material in the vicinity of the neutral axis.

Beams of Uniform Strength.

For most arrangements of the load on a beam, the bending moment at various points of the length varies. It is not generally necessary, therefore, that the section of the beam should be uniform from end to end, and material may be economised by reducing the section where the bending action is less. When a beam is so proportioned that the moment of resistance of the section, at each point of its length, is proportional to the bending moment at that point, the beam is called a beam of uniform strength.

It will only be necessary to consider two kinds of loading—that in which the load is concentrated at one point of the beam, and that in which the load is uniformly distributed over the beam. In the former case, the bending moment is greatest at the place where the load is applied, and in the latter case, the bending moment is greatest at the centre of the beam. In both cases there is no bending moment over the supports. There are also two ways in which the section of the beam may be varied so as to fulfil the conditions of uniform strength. The breadth of the beam may be maintained constant, and the depth varied, or the depth may be maintained constant, and the breadth varied. If the beam is a flanged beam, the thickness of the flanges is supposed to be constant throughout the beam in either case. For a beam of uniform strength and uniform depth, carrying a single fixed load, the plan of the flanges should be two triangles, with their bases united in a line, passing through the point at which the weight is applied and their apices at the supports. But if the flanges are of uniform breadth, then the side-view of the girder should be a triangle, the base of which is the top flange, and the apex of which is directly under the load. For a beam of uniform strength and uniform depth, carrying a uniformly distributed load, the plan of the flanges should be formed by the overlap of two parabolas, whose vertices are at the ends of a line drawn across the centre of the girder, equal in length to the breadth of the flange. But, if the breadth of the flanges is uniform, then the depth must be variable, and one flange may be straight and the other curved in the side-view. The curved flange of the girder should be a parabola whose axis is vertical, and its vertex at the centre of the girder. This form should also be adopted for a girder which is to carry a single moving load, such as the gantry to carry the crab of a travelling crane, and for the girders of bridges to carry weights upon a single line of rails.

Flanged cantilevers (if intended to be of uniform strength) should be of exactly the same form as the half of one of the above girders, according to the manner in which they are loaded. There are often, however, practical difficulties in the way of making beams of the above forms, and they are frequently made half as deep at the ends as in the middle, so as to obtain an abutment for fixing, and they are curved so as to include a parabola drawn through the three points thus determined.

In order that the student may readily ascertain the strength and deflection of any rectangular or cylindrical beam, it is necessary, as already stated, owing to the defective state of our knowledge of this part of the subject, to use certain numbers called constants, which have been determined at various times by different persons.

Practically speaking, there is a great objection to the use of such numbers, but until the position of the neutral axis of such beams as are employed in practice is accurately determined, it is the only course open; and if an engineer has to make a structure of some new sort of timber or other material, upon which no previous experiments have been made, he will have to make experiments upon model beams of the material, to determine the constant number, before he can proceed with any proper feeling of certainty that, in his structure, he has obtained the requisite strength with the minimum quantity of material.

The following table contains, in the column marked 'Value of Strength,' the constants required for ascertaining the ultimate strength of a rectangular beam, of a given size, of most of the materials commonly used in structures; and the strength or breaking weight in lbs. of any such beam, when supported at each end and loaded at the middle, may be found by multiplying the breadth in inches by the depth in inches squared, and by the constant number in the column 'Strength,' and dividing the product by the length of the beam in feet.

The value of the constants for strength may be found, by taking a bar one inch square, twelve inches between the supports, and observing how many pounds' weight applied in the middle is required to break it; the number of pounds so ascertained is the constant. But the experimental beams need not be of the above dimensions; any convenient size of beam may be used for the experiment, the larger the better, and the breaking weight of the experimental beam is then reduced to the weight required to break a beam of the above dimensions.

For example, the constant for the strength of teak, contained in the table, was determined from three experiments made by Barlow upon beams of that material, seven feet long by two inches square, which broke with a mean weight of 938 lbs.

It has been already stated, that the breaking weight of a rectangular beam varies as the breadth and the square of the depth, and inversely as the length, or thus:

$$\text{Breaking weight in lbs.} = \frac{\text{breadth in inches} \times \text{depth squared in inches} \times \text{constant}}{\text{length in feet}};$$

and all these particulars are furnished by the experiment, except the constant, which may be ascertained by the inversion of the above statement; thus:

$$\text{Constant} = \frac{\text{length in feet} \times \text{weight in lbs.}}{\text{breadth in inches} \times \text{depth in inches squared.}}$$

Beams and Girders.

Then, by filling in the known quantities and working out the sum, we obtain the constant; thus:

Constant $= \dfrac{7 \times 938}{2 \times 4} = 820$, which is the number placed opposite teak in the table.

To find the breaking weight of a beam of any of the materials in table:

$$\dfrac{\text{Breadth in inches} \times \text{depth in inches squared} \times \text{constant}}{\text{length in feet}} = \text{breaking weight in lbs.}$$

For example, find the breaking weight at the centre of a beam of Memel deal, 14 inches deep, 10 inches wide, and 20 feet between the supports:

$$\dfrac{10 \times 14 \times 14 \times 545}{20} = 53410 \text{ lbs.}$$

To find the breadth of beam required, so that it will just break with a given weight, when the depth and length between supports are given:

$$\dfrac{\text{Length in feet} \times \text{breaking weight in lbs.}}{\text{depth in inches squared} \times \text{constant}} = \text{breadth required in inches.}$$

To find the depth of beam required to support a given weight, when the breadth and length between supports are given:

$$\dfrac{\text{Length in feet} \times \text{breaking weight in lbs.}}{\text{breadth in inches} \times \text{constant}} = \text{depth in inches squared.}$$

And the square root of the result will be the depth required in inches.

For example, find the depth of a beam of English oak required to support $2\frac{1}{2}$ tons at the centre with safety, to be 10 inches broad and 25 feet between the supports. In this case we must first determine how many times the breaking weight of the beam is to exceed the working load; in permanent structures it should be ten times, but in temporary constructions it may be reduced to six times. In this ex-

On the Strength of Structures.

ample we will take the former, and then the breaking strength becomes—

$$2\tfrac{1}{2} \times 10 = 25 \text{ tons, and } \frac{25 \times 25 \times 2240}{10 \times 557} = 251\cdot 3;$$

and the square root of $251\cdot 3 = 15\cdot 8$, the depth, in inches, of beam required to meet the above condition.

Constants of Strength and Deflection.

The latter will be explained hereafter.

TABLE XXXIX.

Nature of Material.	Value of Strength.	Value of Deflection.	Authorities.
Teak .	820	5588	P. Barlow.
English Oak	557	3359	
Canadian Oak	588	4964	
Dantzic Oak	485	2757	
Adriatic Oak	461	2255	
Ash	675	3807	
Beech	518	3133	
Elm	337	1620	
Pitch Pine	544	2837	
Red Pine	447	4259	
New England Fir	367	5072	
Riga Fir	369	3079	
Mar Forest Fir	381	2013	
Larch	284	2437	
Norway Spar	491	3374	
Mahogany, Spanish	425	2906	Tredgold.
,, Honduras	637	3571	
Memel Deal	545	4500	
Christiana Deal	686	4176	
Cast Iron	2548	41740	Banks.
Do.	2532		
Wrought Iron, Swedish	3473	64221	Kircaldy.
Hammered Steel	6403	78822	

The relative strength of beams of rectangular or other sections, supported and loaded in any other manner, will be

Beams and Girders. 183

to that found by the foregoing rules, in the various proportions given at page 166.

With the flanged girder, the case is widely different, for in its construction we have the means of placing the material (whether of wrought or cast iron) in the most advantageous position, to resist the strain, and, hence, it would be a deliberate waste of material, to make a beam of either of those substances of the form employed for a timber beam. Having this power, we place the bulk of the material at the greatest distance from the neutral axis.

In designing flanged beams, the nature and amount of the strain on each part of the flanges should first be ascertained, and then, if the proper amount of material to resist the strain is introduced at each point, the greatest economy of material will be attained.

It has been previously explained that the top flange will be compressed, and the bottom flange extended, or, in other words, that the material in the top flange will be subjected to a compressive, and that in the bottom flange to a tensile stress. The amount of these strains can be found by considering each half of the beam A B C, in fig. 22, as a bent lever, the long arm being equal to the distance between the weight and the support—namely, in this case, ten feet—and the short arm to the depth of the beam, or in this case two feet. The force acting at the end of the long arm will be the upward resistance of the support, in this case equal to half the weight, or six tons ; and by the principle of the lever, if we multiply the weight acting at the end of the long arm by its length, and divide the product by the length of the short arm, we obtain the weight that will be required at the end of the short arm of the lever to balance the upward resistance of the support. Thus, in the case shown, 10 multiplied by 6, and the product divided by $2 = 30$ tons, the total strain on one flange.

The total amount of strain on the top flange is the same as that on the bottom flange, as indicated in fig. 23.

184 *On the Strength of Structures.*

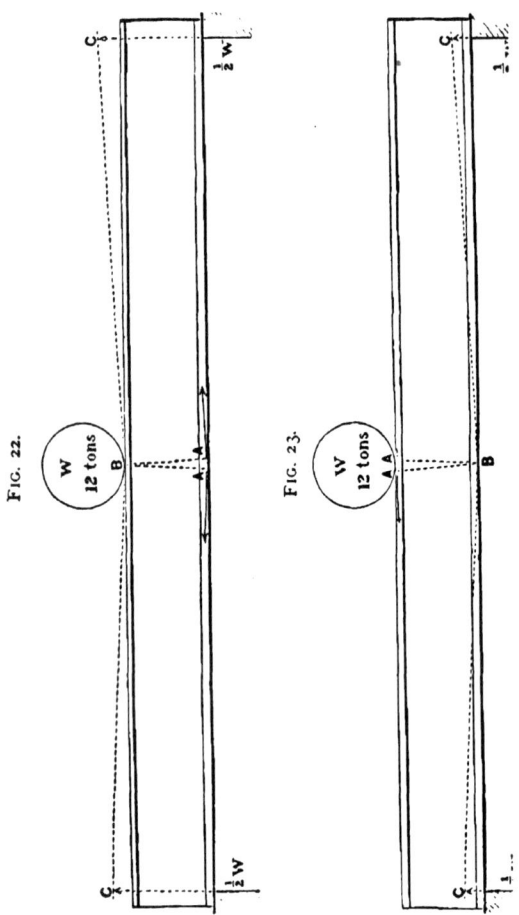

Fig. 22.

Fig. 23.

But the strain on the top flange is compressive, and that the bottom flange is tensile. Supposing that the greatest intensity of stress in tension, for the material of the beam, to be $1\frac{1}{2}$ ton per square inch, the bottom flange must have $30 \div 1\frac{1}{2} = 20$ square inches of sectional area. It may be 10 inches wide and 2 inches thick, for instance, or of any other dimensions giving the same area.

Professor Hodgkinson found, by a number of experiments, that the top flange of a cast-iron girder only required to be $\frac{1}{6}$th of the area of the bottom, or, in other words, that cast iron might be subjected to a compressive stress of nine tons per square inch, but yet we find in practice that the top flange is seldom made less than $\frac{1}{4}$th of the area of the bottom flange. The reason is that, if the top flange is made only $\frac{1}{6}$th of the area of the bottom flange, and of sufficient width to give the required lateral stiffness to the upper part of the beam, it must be very much thinner than any other part of the girder, which would induce an initial strain upon some part of the girder by unequal contraction, due to the different rate of cooling after being cast.

Take, for example, the top flange of the girder in question. If it were made $\frac{1}{6}$th of the area of the bottom flange, it would only contain $20 \div 6 = 3\frac{1}{3}$ square inches; and if it were made 4 inches wide, then it would only be $3\frac{1}{3} \div 4 = \frac{5}{6}$ths of an inch thick, whereas the bottom flange would be 2 inches and the web $1\frac{1}{4}$ inch thick; but by making the top flange $\frac{1}{4}$th of the area of the bottom its section would be $20 \div 4 = 5$ square inches, and if made 4 inches wide it would be $5 \div 4 = 1\frac{1}{4}$ inch thick; that is, equal to the thickness of the web, a much better proportion. Some experience is necessary, to determine the best form and dimensions, so as to meet not only the theoretical requirements of strength, but the conditions imposed by practical experience in the foundry.

By assuming a compressive stress of 6 tons per square inch of section, an area of $30 \div 6 = 5$ square inches will be required to produce the same result, and the flange may be made 4 inches wide and $1\frac{1}{4}$ inch thick, or 5 inches wide and 1 inch thick; but the former would be preferable, because the cooling in the mould would be more uniform, and therefore the contraction of the top flange would be more nearly equal to that of the other parts of the girder.

If the girder is to be of wrought iron, a similar method is adopted, but, with that material, the strain per square inch which is generally allowed is 4 tons for compression and 5 tons for tension, and, according to the Board of Trade regulations, it is 5 tons per square inch both for tension and compression, which simplifies the calculation, is easily remembered, and is sufficiently near for ordinary purposes. At the same time, it would not be applicable to structures exposed to contingencies, such as the surging of slings by falling weights.

Deflection of Beams.

When a beam is supported at each end, the distance to which the middle of the beam is forced down below its original position, by the load, is termed its deflection, and in parallel girders, with the flanges of uniform strength, the deflection curve is found to be circular.

The deflection of solid rectangular beams varies directly as the load and the cube of the length, and inversely as the breadth and the cube of the depth.

In flanged girders, the amount of deflection varies directly as the load, the sum of the areas of both flanges, and the cube of the length, and inversely as the product of the areas of the two flanges multiplied together and the square of the depth of the web.

It appears from the reasoning and experiments of Professor Barlow, that the deflection of a rectangular beam, fixed at one end and loaded at the other, is equal to that of a beam of twice the length, supported at both ends and loaded at the centre with double the weight. The deflection varies as the cube of the length, and if we reduce the girder until it is equal in length to the semi-girder, the ratio of the deflections will then be as the cubes of the lengths; namely, as 1 is to 8.

Beams and Girders.

In other words, the deflection of a semi-girder will be eight times that of a girder, equal to it in all respects, when the latter carries double the load. If we now reduce the weight upon this shortened girder—say ½—so as to make it equal to that carried by the semi-girder, then we reduce its deflection ½, and hence the amount of deflection of the semi-girder will be sixteen times that of the girder when they are equally loaded. It has already been shown that the strength of the girder is four times that of the semi-girder, whereas from the above reasoning the stiffness of the girder is shown to be sixteen times as great as that of the semi-girder.

If we require two beams of the same breadth, but of different lengths, to be equal in stiffness, then their respective depths must be in proportion to their lengths, because deflection, or want of stiffness, varies directly as the cube of the length, and inversely as the cube of the depth, or as the cubes of both these dimensions.

For example, let these lengths be 24 and 12 feet; then if the latter is 12 inches deep, the former will have to be 24 inches deep to be equally rigid, whereas it would be equally strong if made 17 inches deep.

If the beams are equal in depth, but of different lengths, and are required to be equal in stiffness, then their breadths must be as the cubes of the lengths. Taking the same lengths as before—24 and 12 feet—the breadths would have to be in the ratio of 24-cubed to 12-cubed; that is, as 13824 is to 1728, or as 8 is to 1. In other words, the long beam would have to be eight times as broad as the shorter one to be equally rigid, whereas it only requires to be twice as broad to be equally strong.

We have already seen that the strength of a rectangular beam varies as the square of the depth, multiplied by the breadth, and divided by the length, but the stiffness varies as the cube of the depth multiplied by the breadth and divided by the cube of the length.

The stiffness of cylindrical beams varies as the fourth

188   *On the Strength of Structures.*

power of the diameter, and inversely as the cube of the length.

The deflection of similar girders also varies with the manner of loading, and if the load is uniformly distributed over a beam supported at both ends, the deflection will only be ⅝ths of that of the same beam, loaded with the same weight collected at the centre of its length ; and the deflection of a semi-girder uniformly loaded is only ⅜ths of the deflection caused by the same weight acting at the end.

The foregoing investigations and results of experiments upon the deflection of beams are true, provided the visible limit of elasticity of the material is not exceeded, but they are not true of the ultimate deflection, because the law of deflection is very uncertain after the elasticity of the material has ceased to be sensibly perfect, and this condition is reached long before rupture takes place.

The constant numbers which are used for ascertaining the deflection of rectangular beams are contained in Table XXXIX. page 182, in the column marked 'Value of Deflection,' and these constant numbers are deduced from the experiments in the following manner :

The deflection, in inches, of rectangular beams supported at both ends and loaded at the centre is equal to—

$$\text{Deflection in inches} = \frac{\text{the cube of the length in feet} \times \text{weight on beam in lbs.}}{\text{breadth in inches} \times \text{the cube of the depth in inches} \times \text{constant.}}$$

And all these elements are determined by the size of the beam, the weight applied, and the deflection which took place during the experiment, excepting the constant, which is found by the inversion of the above statement ; thus—

$$\text{Constant D} = \frac{\text{the cube of the length in feet} \times \text{weight on beam in lbs.}}{\text{breadth in inches} \times \text{the cube of the depth in inches} \times \text{deflection in inches.}}$$

Take for example the constant for deflection of teak,

Beams and Girders. 189

which has been determined from the same experiments as those from which the constant for strength was obtained, but with this difference, that the weight upon the beam is not the breaking weight, but the greatest weight the beam bore, while its elasticity remained visibly perfect, and the deflection is that which was caused by that weight. These two quantities are stated by Barlow to have been 300 lbs., and 1·151 inch ; then, filling in the quantities, we have—

$$\text{Constant} = \frac{7^3 \times 300}{2 \times 2^3 \times 1\cdot 151} = \frac{343 \times 300}{2 \times 8 \times 1\cdot 151} = 5588,$$

which is the number opposite teak, in the column headed 'Value of Deflection,' and all the constants D in Table XXXIX. page 182, are calculated in this way, from experiments carried out by different individuals.

The deflection of timber beams should not exceed in practice $\frac{1}{480}$th of their length, and it appears, from Tredgold's experiments on cast iron, that if the deflection of bars of that material exceeds $\frac{1}{130}$th of their length, a permanent set is caused; from Kircaldy's experiments on bars of wrought iron and steel, that a deflection exceeding $\frac{1}{600}$th of the length causes a permanent set, upon bars of those materials.

To find the size of beam, supported at both ends and loaded at the centre, capable of supporting a given weight with a given amount of deflection :

$$\frac{\text{The cube of the length in feet} \times \text{weight in lbs.}}{\text{deflection} \times \text{constant}} = \begin{cases} \text{breadth in inches} \times \text{cube of} \\ \text{depth in inches.} \end{cases}$$

For example, find the size of a beam, of English oak, supported at each end and loaded at the centre with a weight of $2\frac{1}{2}$ tons, the distance between the supports being 25 feet, and the deflection not to exceed $\frac{1}{480}$th of the length ; the deflection in inches will be—

$$25 \times 12 \times \frac{1}{480} = \frac{300}{480} = \frac{5}{8} = \cdot 625 \text{ of an inch;}$$

then $\dfrac{25^3 \times 2\cdot 5 \times 2240}{\cdot 625 \times 3359} = 41774 = \begin{cases} \text{the breadth} \times \text{the cube} \\ \text{of the depth :} \end{cases}$

and, assuming 10 inches for the breadth, we have $\frac{41774}{10} = 4177\cdot 4 =$ the cube of the depth, and the cube root of $4177\cdot 4 = 16\cdot 1 =$ the depth of the beam required.

If the beam is to be square, the fourth root of the quotient will be the side of the square ; thus—

The fourth root of $41774 = 14\cdot 3$ nearly, for the side of the square.

If the beam is to be cylindrical, first multiply the quotient by $1\cdot 7$, and then extract the fourth root, which will be the diameter of the beam required, because the deflection of a cylindrical beam is $1\cdot 7$-ths that of a square beam, all other circumstances being the same, and hence it requires $1\cdot 7$-ths the material to render it equally rigid. The diameter of a circular beam will therefore be the fourth root of

$41774 \times 1\cdot 7 = 16\cdot 3$ inches = the diameter of the beam.

If the beam is to be rectangular, either the breadth or depth must be fixed, and the other dimension will be found thus.

To find the depth of beam required to carry a given weight, with a given amount of deflection, when the length and breadth of the beam are given:

$$\frac{\text{The cube of the length in feet} \times \text{weight in lbs.}}{\text{breadth in inches} \times \text{deflection} \times \text{constant}} = \text{the cube of the depth;}$$

then the cube root of the result is equal to the depth in inches required.

To find the breadth of beam required when the weight to be carried, the amount of deflection, and the length and depth are given:

$$\frac{\text{The cube of the length in feet} \times \text{weight in lbs.}}{\text{the cube of the depth in inches} \times \text{deflection} \times \text{constant.}} = \left\{\begin{array}{l}\text{the breadth in inches} \\ \text{required.}\end{array}\right.$$

Beams and Girders. 191

To find the weight which will cause a given amount of deflection upon a beam:

$$\frac{\text{Breadth in inches} \times \text{the cube of the depth in inches} \times \text{deflection} \times \text{constant.}}{\text{the cube of the length in feet}} = \text{weight in lbs.}$$

To find the deflection of a beam supported at both ends with a load uniformly distributed over its entire length, take ⅝ths of the result, given by the rule to find the deflection of a beam loaded at the centre.

To find the deflection of a semi-beam, supported at one end and loaded at the other, multiply by 16 the deflection of the same beam supported at both ends and loaded at the centre.

And, lastly, to find the deflection of a semi-beam, supported at one end, with the load uniformly distributed over its entire length, multiply by 6 the deflection of the same beam supported at both ends and loaded at the centre.

The relative deflection of similar beams, each supporting the same weight, with the supports and load in various positions is, as follows:

Position of Support and Load.

	Relative deflection.
When supported at both ends and the loads evenly distributed, the deflection is	5
When supported at both ends and loaded at the centre, the deflection is	8
When supported at one end and load distributed, the deflection is	48
And when supported at one end and loaded at the other, the deflection is	128

The above table teaches a very instructive lesson to engineers, and shows how wrong in principle it is, to have wheels, pinions, or pulleys, which have hard work to perform, overhanging the bearing upon which they are supported, and in the case of machines, where such an arrangement is rendered necessary, the shaft should then be made of proportionately increased diameter in the bearing, and tapered off, being not abruptly, but gradually diminished to the diameter of the remaining portion of the shaft.

These relative deflections of the beam and semi-beam, loaded in different ways, apply equally to beams of any section, so that if the deflection of the simple girder, supported at each end and loaded at the centre, be found, that of the others can be ascertained by simple multiplication, as above stated.

If the student requires to find the deflection of a beam of any other form of section than the rectangular or circular, he must first deduce the constant for that particular form of section, from an experiment upon a similar beam, and then proceed in a similar manner, or else he may make an experiment on his own account, which will be useful to himself in other respects.

Resilience of Beams.

The resistance of beams to transverse impact, or, in other words, to a suddenly applied load, is termed their 'resilience,' and it follows a very different law from that of their strength, for it is simply proportional to the mass or weight of the beam, irrespective of the length, or the proportion between the depth and breadth.

It appears from the published experiments and statements of the Railway Commissioners, that a beam 12 feet long will only support $\frac{1}{2}$ of the steady load that a beam 6 feet long of the same breadth and depth will support, but that it will bear double the weight, suddenly applied, as in the case of a weight falling upon it; or if the same weights are used, the longer beam will not break by the weight falling upon it, unless it falls through twice the distance required to fracture the shorter beam.

This law was apparently proved by a large number of

On the Strength of Gearing. 193

experiments carried out for the Railway Commissioners, in which the beams appear to resist the sudden application of the load, by gradually absorbing the work accumulated in the falling weight, and to bend through a certain distance, until the work done in bending the beam through that distance is equal to the work accumulated in the weight, due to the distance which it has fallen. Hence a very strong beam may be broken, if it is not sufficiently elastic to bend through the required distance, and thus absorb the *vis viva* of the falling weight.

CHAPTER XIII.

ON THE STRENGTH OF GEARING.

IN the chapter on torsion, reference is made to the strength of spindles and shafts, employed in conveying power from one point to another, to give motion to machinery. It is equally important for the engineer to know, how to proportion the other parts of the transmissive machinery—the spur-wheels and bevil-wheels, for instance—which are the agents by which power is transferred from one shaft or spindle to another. More especially is it important, to be able to calculate the strength of the teeth of wheels, these being the agents through which the driving forces are directly transmitted.

When properly made, the teeth of the two wheels act against each other, as the wheels revolve, with comparatively little thrust, noise, or friction. To secure proper mutual action of the teeth, their form must be determined on the principles which are explained in Professor Goodeve's 'Treatise on Mechanism.' But, besides determining the best form for the teeth, the engineer requires, also, to ascertain their size for any given case, and this is usually simply a question of strength.

In estimating the strength of the teeth of mill-gearing, we have to attend, first, to the strength of the material of which the gearing is made ; second, to the forces which act on the teeth, due to the power transmitted ; and, third, to the way in which the teeth resist fracture under the action of the ascertained forces. We shall best explain the method of proceeding, by taking an example. Suppose it is required to determine the dimensions of the teeth of a wheel on the main axle of a thirty-ton crane. Let the barrel be four feet diameter, and the wheel six feet diameter, measured to the pitch-line. Further, let it be assumed that the tension on the chain which is coiled on the barrel is reduced from 30 tons to 7½ tons, as the result of the mechanical advantage of the chain tackle. The total load on the teeth of the wheel will be 7½ tons, multiplied by the radius of the barrel, and divided by the radius of the wheel ; that is, 7½ × 2 ÷ 3 = 5 tons, or 11,200 lbs. We may presume, that this load is distributed over two teeth, there being always at least two teeth in full bearing at the same time. The load on one tooth is therefore 11,200 ÷ 2 = 5,600 lbs., which acts on the tooth like a load on the end of a projecting cantilever, tending to break it by transverse fracture at the root.

Let it be next assumed that the wheel is of cast iron, and that, for safety, the strain on the tooth is not to exceed $\frac{1}{15}$th of that which would fracture it. A cast-iron bar of good quality, 1 inch long and 1 inch square, loaded at the end, would break with about 6,000 lbs. The wheel tooth is to be considered, as in similar conditions to such a bar, or, in other words, it is in the condition of a beam loaded at one end and fixed at the other, and its resistance to fracture is proportional to the square of its depth (that is, the thickness of the tooth) to the breadth (that is, the width of the face of the wheel) and inversely as the length (that is, the projection of the tooth measured from the root to the point).

As the length and thickness of the teeth of wheels are

usually in some definite proportion to the pitch, it will now be necessary for the student to assume, to the best of his judgment, a suitable pitch of tooth for the required purpose, which will give the length and thickness. For example, assuming the pitch to be 2·5 inches, the length will be 1·875 inch, and the thickness at the root 1·55 inch, the breadth across the wheel being three times the pitch, or 7·5 inches. Squaring the depth of the tooth, multiplying by the breadth and the strength of the iron, and then dividing the product by the length of the tooth, we get—

$$\frac{1·55^2 \times 7·5 \times 6000}{1·875} = 57660 \text{ lbs.},$$

the ultimate strength of one tooth; and the double of that, namely, 115,320 lbs. for the two teeth, that are supposed to be in gear. The ratio of this to the strain upon the teeth will therefore be $\frac{115320}{11200}$, or 10·3, the ratio of the strength to the working load. Should the assumed size be found either too strong or too weak, the assumed pitch must be decreased or increased until a suitable tooth is discovered

In practice, the strength will much depend upon the accuracy of the adjustment of the gear, for if the teeth bear only at one end—a condition frequently seen—then the great advantage to be derived from a broad wheel is necessarily lost, and might as well not have been provided. In addition to proper adjustment, a great increase of strength is gained by flanging the teeth of both wheel and pinion up to the pitch-lines; this arrangement is now generally adopted for wheels of importance.

In determining the substance to be given to the teeth of wheels to afford a given amount of strength, it has to be kept in view that such wheels, and the pinions especially, are subjected to rapid wear when in daily use, thus reducing the thickness of the teeth; in other words, the depth of the beam and its strength, which is as the square of its depth. Hence it is necessary to make some allowance for the future wear in the strength of the original construction.

From the circumstance that the pinion, in a given period, revolves so many times oftener than the wheel, the usual custom of making the teeth of both with the same allowance for wear may seem to be incorrect, and so far that is the case; but still, as a rule, there are practical objections in the way of the more correct arrangement, which prevent its general application, although for special purposes it is sometimes otherwise. Thus, for gunpowder machinery, where great pains are taken, and where gun-metal and hornbeam work together, it is usual to reduce the metal and add a corresponding extra thickness to the wood.

In the foregoing remarks, cast iron is chiefly referred to, because it is used more than any other metal for wheel purposes. But that material is not selected on account of its special fitness or superiority in any respect, so far as the duty to be performed is concerned; it is chosen more on account of its cheapness, and because it may be readily cast into any form of wheel. Such considerations have great influence in the settlement of such points, and particularly so in this case, for, owing to the peculiar form given to spur and bevil wheels, it is much easier to cast them than to forge them; indeed, to make large toothed wheels by forging is scarcely practicable, on account of the cost. For smaller wheels, termed pinions, and where a great number are required of a definite size, it is perfectly easy to make them by forging. Simple dies are prepared, in which a roughly shaped, viscous mass of white-hot wrought iron is placed, and then subjected to the blows of a steam-hammer. This, together with the aid derived from a taper punch, driven through the centre, forces the soft iron into every crevice of the die. By such means, wrought iron is used for wheels to some extent, and the dies may be so formed that the teeth will be flanged up to the pitch-lines, and pinions so made are at least three times stronger than the same forms when made of cast iron.

In circumstances where only one pinion or a few pinions, made of wrought iron, are required, it is more convenient (in order to avoid the cost of the dies) to forge them into solid blocks, and then cut out the teeth afterwards. This arrangement, however, scarcely admits of flanges, on account of the cost of manufacture, and the want of flanges may be said to reduce the strength by one-third; consequently, such wheels have only double the strength of cast-iron pinions or wheels, made with flanges.

Of late years, the material called malleable cast iron has been employed for many such purposes; that is to say, the wheels are made of cast iron in the ordinary manner, and then subjected to a course of annealing, while embedded in some substance rich in oxygen, which combines with a portion of the carbon in the iron, and leaves the metal malleable. This process is the reverse of steel-making by cementation. Wheels made by this method may have flanges, and have double the strength of cast iron, and are used extensively where great accuracy is not essential, but it is found difficult to maintain the truth of the pitch and form of the teeth, during the annealing process. During this process the casting is kept for a long time at a red heat, and this frequently causes it to twist or warp.

All difficulty is, however, now removed by the use of cast steel, which may be cast into any form in an earthen mould, in the liquid state, in the same manner as cast iron, and, when properly made, such cast-steel wheels have four times the strength of cast-iron wheels, the only objection being the cost, cast steel in this form is nearly as expensive as bronze.

The amount of mechanical power that may be transmitted by a pair of wheels, depends entirely upon the speed at which they are driven; this opens up another and entirely different question, from that of the strength of the teeth of a wheel for a crane. In ordinary mill-gearing a wheel that might be capable of transmitting 10 horses' power at 60 revolutions per minute, will be able to transmit 20 horses' power, if it is driven at 120 revolutions; the same remarks apply to all the other parts, such as pulleys, and every kind of moving mechanism. But, practically, this is only true within certain limits, for, on reaching high velocities, other conditions of impact, vibration, and centrifugal force due to velocity step in, to keep the mechanical application within bounds, as depending on the material which is employed for the construction.

Strength of Screws.

A little consideration will show the student that the strength of the thread of a screw-bolt, or of its nut, or that of the tooth of a tangent wheel, depends on principles similar to those applicable to ordinary gearing. Looking at the longitudinal section of a portion of a screw, the thread will be found under the same conditions as a beam fixed at one end and with the load distributed uniformly, or precisely similar to that of the tooth of a spur-wheel; if, for example, the screw thread is made of a square form, it will have less breadth of root, and consequently less strength, than the same pitched screw, when made in the usual conical form, or even when made with the top and bottom rounded off, inasmuch as both of the latter shapes have a broader base to be broken, or even detruded, thus affording a better resistance.

Although the common angular thread, in one respect, may be considered as the strongest form, still it is found that, in another respect, it is not so strong as the square thread—namely, in resisting the influence of wear, as arising from the inclined plane action of the angle upon a corresponding angular surface in the nuts—which is remedied in the flat bearing of the square thread. Consequently the square thread is found, in practice, to be not nearly so likely to override the nut, by any excessive wear or by any inordinate straining.

On the Strength of Gearing. 199

Screw threads with a round top and round bottom are intended to combine both advantages—to possess the broad base of the angular thread, and at the same time to have some portion of the flat-bearing surface of the square thread. By this compromise they have, upon the whole, a decided advantage, which gives them the preference for many purposes.

As a rule, screws act in one direction only; such is the case with screw-bolts and nuts; but there are many exceptions, such as the leading screw of a lathe or planing machine, or wherever the screw is employed as an agent to impart motion; screws for the latter class of purposes are correctly formed when both sides of the thread are alike, because both sides have to perform the same duty. It is different, however, with the majority of screws, and especially so in the example given—namely, the bolt and nut. There is no mechanical or manufacturing reason, why such screws should not be made at right angles to the axis, upon one side of the thread, so as to have the bearing surface similarly placed to that of a square thread, and with the other side inclined, as in the ordinary triangular thread. The threads would then be, in section, similar in form to the tooth of a ratchet-wheel, which acts only in one direction. Such a form has a self-evident advantage, in the case of the ratchet-wheel, but it would have a still greater advantage in the screw, because it would not only give the same breadth of base as the common angular thread, but that advantage would be combined with the complete flat bearing surface of the square thread, or, in other words, it is a form possessing the full advantage of both, and without any of their disadvantages.

This form of screw thread was adopted by Sir William Armstrong, for the breech-screws of his celebrated guns, and its superiority in strength over the square threaded screw, formerly used, was so conspicuous as to leave no doubt in regard to its comparative advantage; the form is correct, in principle, for screws that act only in one direction, and, as the arts advance, must become general.

Strength of Cutting Instruments.

Similar principles might be advantageously applied in many other cases, but they are frequently lost sight of, and, although a great change for the better has taken place, of late years, more especially in regard to the forms given to cutting instruments of every description, so as to combine great strength with incisive penetration, we are yet far from having the minds of our workmen so imbued with the natural principles involved, as to keep them right, by an intuitive perception of what that right is, due to the natural reason which guides us in so many other things, and which is independent of men's inventions, and is the effect of time and training.

To select a cutting instrument for the lathe or planing machine, as an example ; when such an article is formed on correct principles, with the side under the cutting edge nearly perpendicular to the work upon which it has to bear, in order to support the cutting edge, and to give great strength, and with the other side bevilled off from the cutting edge, like the saw-tooth, so as to give the knife-action of penetration ; such an instrument will be immensely more efficient, and many times more economical, than when the form given to the same instrument has been devised without knowledge and with reference to penetration only. Forms which are good in the estimation of the workman, have frequently neither strength nor penetration, the metal of the instrument being ground away where it should be left, and being left at the part where it should be ground away.

Such a lack of proper conditions arises, entirely, from not knowing the true principles, as there is no difference in cost or trouble. The result is a want of strength, which incapacitates the instrument from grappling with the work, and

On Columns. 201

tiny shavings are peeled off by it, in a broken condition and in miserable quantity. An iron shaving of one continuous curl was shown in the 1862 Exhibition, 1140 feet in length, the length of the curl being 462 feet; and some of the Royal Gun Factory iron shavings are $\frac{1}{4}$ of an inch in thickness by $4\frac{1}{2}$ inches in breadth, and are curled up of any length, as if they were wooden spills. All this efficiency is due to the form which is given to the cutting instrument, which of course has to be supported by a machine of corresponding strength and power, but the perfection of cutting is entirely owing to the application of the true principles of strength and penetration, in shaping the small piece of cast steel which is used as an instrument. The mass of iron in the shaving indicates the strength of the tool, and the unbroken condition of the curl is an evidence of its fitness for the removal of the superfluous material, and at the same time shows that it is accomplished, without that useless expenditure of mechanical power, which is the case with more detrusive instruments. With badly formed tools, the shaving is broken up into small chips, to no useful purpose. Thus it is shown that 'knowledge is power.'

CHAPTER XIV.

ON THE STRENGTH OF LONG COLUMNS.

THE resistance of a long column, to loads acting in the direction of its axis, depends mainly on three conditions : first, on the proportion which its length bears to its least transverse dimension ; second, on the form of the ends of the column ; and, third, on the direction in which the load acts, with reference to the axis of the column.

In the case of short columns—that is, columns whose

On the Strength of Structures.

length is only slightly in excess of their transverse dimensions—the material is ruptured by simple crushing alone; but when the height exceeds from three to eight times the least transverse dimension, according to the nature of the material, the rupture is caused partly by bending and partly by crushing; and, when the length exceeds from twenty-five to thirty times the transverse dimension, then the column will fail by bending, and the material will be subjected to strains, similar to those of a beam, when under a transverse load; one side will be crushed and the other will be extended.

There is, at the present time, no completely satisfactory theory of the ultimate resistance of long columns. Euler investigated the law of resistance, on the assumption that the elasticity of the material remained perfect up to the point at which rupture was imminent. On this assumption, he found that the resistance of long cylindrical columns would be proportional to the fourth power of the diameter and inversely as the square of the length.

Hodgkinson found, however, by his experiments, that the ultimate resistance was proportional to a power of the diameter rather less than the fourth, and decreased in a much less ratio than the square of the length.

Table XL. p. 203, gives the mean results of a number of experiments, made by Hodgkinson, and shows clearly that the strength of pillars depends greatly on the secure fixing of their ends. With both ends rounded, the strength is only $\frac{1}{3}$ of that afforded by similar pillars, having both ends flat, and abutting on flat surfaces. Pillars, with one end round and the other flat, have $\frac{2}{3}$ of the strength of those with both ends flat. These facts show the advantage derived from correct bearing surfaces in structures exposed to compression. The table also shows that pillars of timber, when the length exceeds seventeen times the diameter, are destroyed by bending, and that even those of seventeen diameters are partly bent as well as crushed.

Results of Experiments made by Hodgkinson to ascertain the Ultimate Strength of Pillars of Timber.

TABLE XL.

Name of timber.	Description of pillar.	Length in inches.	Sectional dimensions in inches.	Ratio of length to least side.	Mean weight in lbs. under which the pillars sank	Remarks
Dantzic oak.	Uniform with both ends rounded.	60.5	1.75 square	34.5 to 1	3,197	Bent and broke.
	Uniform with one end round and the other flat.	60.5		34.5 to 1	6,109	Round end crushed.
	Uniform with both ends flat.	60.5		34.5 to 1	9,625	Sank by bending.
		29.7		17 to 1	13,083	Crushed at the ends.
		30.25		17.3 to 1	14,305	Wrinkled at the ends, bent and sank.
Red deal.		48	1 square	27.4 to 1	9,229	Bent and sank.
		46	1.5 square	46 to 1	1,754	Bent without crushing at the ends.
		58	2 square	37 to 1	7,888	
		58	2 square	29 to 1	12,385	One end crushed—remaining two-thirds of column cripped.
		58	2.9 × 1.4	29 to 1	11,601	
		58		43 to 1	7,681	Sank by bending, in the direction of smaller side.
		58	3.47 × 1.15	50 to 1	4,349	

It was found in these experiments that a deflection was visible, in one case, with a little under $\frac{1}{5}$ of the destroying load, and that, generally, there was a considerable deflection with between $\frac{1}{3}$ and $\frac{1}{4}$ of the destroying load.

The visible result of the crushing strain upon short columns varies with the nature of the material, and rupture is caused either by splitting, shearing, or bulging; the former is characteristic of the hardest cast iron, the hardest description of stones, and also of timber. Crushing by shearing is exhibited by cast iron, when the height of the column is about one and a half time the transverse dimension, whilst crushing by bulging takes place with short specimens of wrought iron, mild steel, gun-metal, lead, and other ductile metals.

There are two other forms of crushing: first, crushing by wrinkling up, commonly called 'crippling' or 'buckling;' and, second, by cross-breaking. The former is sometimes seen in columns of wrought iron, which are too long to be crushed by bulging, and too short to be bent by flexure; and the latter occurs in cast-iron columns, when the length exceeds thirty times their diameter.

The manner in which the column gives way depends upon the form of the ends of the column, in so far that, if the ends are rounded, the column is as liable to flexure as one of the same diameter and twice the length, with both ends flat and firmly fixed. Hence, if we break two similar columns, the one with the ends rounded, and the other with the ends flat, the length being about twenty times the diameter, the material in the one with rounded ends will be ruptured entirely by cross-breaking or transverse strain, while that of the other will be ruptured partly by crushing and partly by cross-breaking.

The way in which the column yields also depends upon the direction of the load, which has heretofore been supposed to pass along the axis of the column. If the direction of the action of the load forms only a very small angle with the axis of the column, it induces a strain upon the column at right angles to the direction of its axis, and it is then in a precisely similar condition to a beam supported at each end and loaded at the centre.

In short columns, possessing sufficient rigidity to resist the bending strain caused by this indirect action of the load, the material is liable to be ruptured in detail, from the fact that the whole of it has not the opportunity of taking its full share of the work in resisting the load, and therefore cannot give due support to the smaller portion, upon which the load acts.

There are a few practical deductions to be drawn from these three considerations : First, that a column should be made as short as possible, in proportion to diameter; in other words, it should be capable of maintaining itself vertical by its stiffness. Second, that the ends of cast-iron pillars should be cast with broad bracketed flanges, considerably larger than the transverse dimensions of the columns at the centre, for although the flanges may not add to the strength, in a direct manner, yet they certainly add to the stability, and will prevent the column from bending ; that is, if the ends are made sufficiently stiff and strong to resist the thrust. Third, that the ends should be at right angles to the axis of the column, and placed so that the load may act upon the whole surface of the capital and base, and its resultant may pass directly along the axis of the column.

As a matter of economy, all long columns should be of the form recommended for cylindrical beams—namely, that of a parabolic spindle—but this is a form which does not please the eye so well as the graceful outline of a taper column, and hence, no doubt, strict economy is often sacrificed to appearance.

In considering the strength of columns, the probability of crushing by splitting or by shearing may be entirely neglected, because, practically, columns are never made so short that these forms of rupture are called into play, and, so far as timber and cast-iron columns are concerned, rupture will always take place by cross-breaking, and wrought-iron plate columns should either be properly stayed, to prevent buckling or bulging, or else sufficient material should be put into the section to resist the stress tending to cause local distortion or wrinkling up of the metal, until the cross-breaking or bending strain can come into play.

The most reliable experiments upon the strength of columns are those of Hodgkinson, who carried out an elaborate series of experiments to determine the laws which govern the strength of cast-iron columns, and who also made a considerable number of experiments for the Railway Commissioners, in order to determine the best form of section of a wrought-iron tube, to resist compression, when the load is applied in the direction of its length.

With respect to cast-iron columns, he found, first, that the strength of solid cylindrical columns with both ends rounded, and the length of which exceeded fifteen times the diameter, varied as the 3·76th power of the diameter; second, that the strength of solid columns with both ends perfectly flat, when the length exceeded thirty times the diameter, varied as the 3·6th power of the diameter nearly; third, that when the diameters remained the same, the strength varied inversely as the 1·7th power of the length; fourth, that the strength of hollow cast-iron columns, with rounded ends when the length exceeded fifteen times the diameter, varied as the difference of the 3·76th power of the external diameter and the 3·76th power of the internal diameter, divided by the 1·7th power of the length; fifth, that the strength of hollow cast-iron columns with both ends perfectly flat, when the length exceeded thirty times the diameter, varied as the difference of the 3·6th power of the external diameter and the 3·6th power of the internal diameter, divided by the 1·7th power of the length; sixth, that the strength of hollow cast-iron columns of the same diameter varied inversely as the 1·7th power of the length, as in the case of solid columns.

On Columns. 207

From the same experiments he deduced the following rules, to assist us in ascertaining the strength of such columns: First, for solid cast-iron columns with both ends flat, when the length exceeds thirty times the diameter, the breaking weight, in tons, equals the product of the 3.76th power of the diameter, in inches, multiplied by 44·16 (a coefficient deduced from the experiments), and divided by the 1·7th power of the length, in feet. Second, for hollow cast-iron columns with both ends flat, when the length exceeds thirty times the diameter, the breaking weight equals the product of the 3·6th power of the external diameter, in inches, less the 3·6th power of the internal diameter, also in inches, multiplied by 44·34 (a coefficient deduced from the experiments) divided by the 1·7th power of the length, in feet.

For columns with both ends rounded, or fixed by pins passing through the ends, ⅓rd of the above strength only should be reckoned upon, and for columns with one end flat and the other end round, ⅔rds of the result by the above rules.

The iron, from which the columns used in the experiments were made, possessed a crushing strength of 49 tons per square inch, and the coefficients should be increased or decreased, in the ratio of that strength, compared with the strength of the iron of which any other columns are made.

It was also found by these experiments that, if columns with flat ends are less than 30 times their diameter in length, or if the ends be rounded and the length is less than 15 times the diameter, that they will be partly crushed as well as bent; and their strength is ascertained as follows:

First ascertain their strength as columns with the ends flat, by the rule for columns exceeding 30 times the diameter in length, or, if the ends are rounded, 15 times the diameter; then their real breaking strength will equal the product of the strength found as above, multiplied by the crushing strength of the material of which the columns are to be made, as ascertained by experiment, divided by the strength found as above plus ¾ of the crushing strength of the material.

From the experiments carried out with solid wrought-iron columns, it appears that the strength of square pillars, which are long enough to be bent before the material is much crushed, varies nearly as the 3·6th power of the side of the square, the lengths being equal.

The results of the experiments upon rectangular tubes of wrought iron are shown in the following Table, from which it appears that, in tubes of equal thickness, the strength per square inch of section of the smaller tubes is greater than that of the larger, the 4-inch square tube, ·06 of an inch thick, giving 8·6 tons per square inch; the rectangular tube, 8 inches × 4 inches, giving 6·79 tons per square inch; and the 8-inch square tube giving 5·9 tons per square inch of section.

Result of Experiments made to ascertain the Resistance to Compression of Rectangular Tubes of Wrought Iron.

TABLE XLI.

Length of tubes in feet and inches.		Form of section.					
		Rectangular 8 inches × 4 inches.		Square 8 inches side.		Square 4 inches side.	
		Thickness of plates in inches.	Strength per square inch of section in tons.	Thickness of plates in inches.	Strength per square inch of section in tons.	Thickness of plates in inches.	Strength per square inch of section in tons.
Feet.	Inches.						
10	0	—	—	—	—	·03	4·9
2	6	—	—	—	—	·03	5·5
10	0	·06	6·79	·06	5·9	·06	8·6
2	4	·06	7·1	—	—	—	—
10	0	—	—	—	—	·083	11·24
2	6	—	—	—	—	·083	12·24
10	0	—	—	·139	9·1	·134	9·63
7	6	—	—	—	—	·134	10·35
10	0	—	—	·219	11·84	—	—
10	0	·264	12·01	—	—	—	—

All these tubes failed by buckling or wrinkling up of the plates.

On Columns.

It will be seen also from Table XLI., that, to obtain twice the strength, four times the thickness of plate had to be used, whereas in some previous experiments on the crushing strength of wrought-iron plates, similar to a side of these tubes, the strength varies as the cube of the thickness.

Result of Experiments made to ascertain the Resistance to Compression of Cylindrical Tubes of Wrought Iron.

TABLE XLII.

Length in feet.	External diameter in inches.	Thickness of plates in inches.	Area of section in square inches.	Ratio of diameter to length.	Ratio of thickness to diameter.	Strength per square inch of section in tons.	Remarks.
10 5 2·5	1·495	·101	·4443	$\frac{1}{80}$ $\frac{1}{40}$ $\frac{1}{20}$	$\frac{1}{15}$	6·55 13·92 15·27	
10 5 2·5	1·964	·104	·6104	$\frac{1}{60}$ $\frac{1}{30}$ $\frac{1}{15}$	$\frac{1}{18}$	10·35 14·86 16·5	All sank by flexure.
10 5 2·5	2·49	·107	·8045	$\frac{1}{48}$ $\frac{1}{24}$ $\frac{1}{12}$	$\frac{1}{23}$	13·29 15·67 16·29	Sank by flexure.
10 2·5	2·35	·242	1·605	$\frac{1}{51}$ $\frac{1}{13}$	$\frac{1}{10}$	9·6 14·78	Both sank by flexure.
10 7·5 2·5	3·0	·151	1·349	$\frac{1}{40}$ $\frac{1}{30}$ $\frac{1}{10}$	$\frac{1}{20}$	12·36 13·3 16·7	Slightly bent.
10· 7·5	4·05	·14	1·7	$\frac{1}{30}$ $\frac{1}{22}$	$\frac{1}{29}$	12·34 14·88	} Crushed.
10 7·5	6·36	·13	2·54	$\frac{1}{19}$ $\frac{1}{15}$	$\frac{1}{48}$	16·02 18·6	} Failed by crippling.

It also appears that, in square tubes compressed to such a high degree, the length has little effect upon the strength, for in these experiments the 10-feet tubes stood nearly as much as the 2 feet 9 inch tubes.

It is obvious, from careful study of these tables, that wrought iron is unsuitable for columns, for in Table XLII. we see that columns only 12, 13, and 15 times the diameter sank by flexure, and in one case, where the length was only 10 times the diameter, the column was bent, and in Table XLI. we find a 4-inch tube 2 feet 6 inches long, or only $7\frac{1}{2}$ times the side of the square, failed by the buckling of the plates; and, in one case, a very short tube (1 foot $7\frac{1}{2}$ inches only), with a section of 8 inches × 4 inches and ·06 of an inch thick, failed in the same manner, although the length was only 5 times the width of the least side. Still, columns are often made of wrought iron, and notably in the case of poles for sheer-legs, sometimes to over 100 feet long, to lift 120 tons and upwards, and by using these tables of strength, and allowing such a margin that the breaking strength shall be 10 times the working load, and stiffening by internal T iron ribs or L irons, the structure will be perfectly safe.

To find the strength of a wrought-iron column, multiply its sectional area in square inches, by the strength per square inch given in these tables, for the column bearing the nearest relative proportion of diameter or side of square to length, and of thickness to diameter or side of square; this will approximately give the strength of the column required.

Steel is used in the present day, to some extent, for the poles of sheer-legs, and will no doubt in time supersede wrought iron for such purposes; for it is equally safe, and combines with its safety a compressive strength superior to either wrought or cast iron, and is not nearly so liable to flexure as the former, nor to fracture as the latter.

Some portable steel sheer-poles were recently constructed for the War Department to carry 18 tons on each pole, or

On Columns.

36 tons altogether, both when in a vertical position and, also, when fixed at such an angle that the top end is 15 feet from the vertical through the bottom end. These poles are 40 feet long, 15 inches in diameter at the centre, and 8 inches at the ends, the thickness of plate being only ·185 of an inch, the ratio of diameter to length being $\frac{1}{32}$ at the centre and $\frac{1}{60}$ at the ends, and the ratio of the thickness to diameter $\frac{1}{81}$ at the centre, and $\frac{1}{43}$ at the ends.

The lower ends of these poles are fitted with ball and socket joints, in order to allow the necessary amount of lateral movement, and the blocks, which carry the weight to be lifted, are suspended from a pin, passing through the upper ends of the poles. Consequently these poles must be treated as columns with rounded ends, and they are only capable of supporting $\frac{1}{3}$ of the weight which similar poles, with ends flat and fixed, would support.

This pair of sheer-poles was tested with a weight of 36 tons, when in the inclined position of 15 feet out of the perpendicular, in which position a load of 36 tons is equal in effect to a load of 40 tons with the poles in the vertical position, and it produces a stress of 2·3 tons per square inch of section at the centre of the poles. This load they withstood, without the slightest perceptible deflection.

It will be seen by reference to Table XLII. page 209, that a cylindrical column of wrought iron with both ends flat, and a diameter equal to $\frac{1}{30}$ of the length, and a thickness equal to $\frac{1}{23}$ of the diameter, was crushed with 12·34 tons per square inch of section.

The crushing strength of a similar wrought-iron column, with both ends rounded, would be $\frac{1}{3}$ of 12·34 tons, or 4·11 tons per square inch of section; and, allowing the safe load of such a column to be $\frac{1}{4}$ of the crushing load, the working strength would be about 1 ton per square inch of section, or less than $\frac{1}{2}$ the stress to which the above steel poles were subjected.

This single example will serve to show that steel may, in

P 2

time, be extensively applied with great advantage, in connection with such structures, and more particularly will this be the case, when they have to be transported from one place to another, and frequently dismounted and re-erected in an entirely different situation, because the whole of the gear, both for their transport and re-erection, would be proportionately lighter.

CHAPTER XV.

ON THE STRENGTH OF CRANES AND ROOF TRUSSES AS EXAMPLES OF COMPLEX STRUCTURES.

WHEN the student has mastered the preceding chapters upon beams and pillars, comparatively little difficulty will be experienced in understanding the present chapter, which treats of the strength of more complex structures. Most of the examples here given are chosen from actual works, which have recently been carried out in the War Department, and although some of them may at first appear difficult to understand, they are really simple, if the principles applicable to them are carefully studied.

Before commencing to ascertain the strength of a structure, the whole question raised should first be broadly and closely considered, in order to determine the direction and nature of the forces to which the structure may be exposed in its various parts. It will then be found that by far the larger proportion of these forces, in modern structures, are in the direction of the length of the parts of the structure, and even the majority of the transverse forces are so supported, that they give rise to simple tensile and compressive strains, which act in the direction of the length of the component parts of the structure. It is comparatively rare to find parts exposed to forces which tend to rupture them by cross-breaking, as in the case of a beam.

On Roof Trusses and Cranes.

This conversion of the direction of the strains is mostly effected, by skilfully forming the structure of one or more triangles or systems of triangulation, or, in more familiar terms, by bracing the component parts of the structure.

The triangular form of structure is now used, in preference to any other, simply because a triangle is the only figure whose shape cannot be altered, while the length of its sides remains constant.

One of the simplest and most familiar forms of a braced structure is that of a common roof truss of small span, as shown in Fig. 24. It consists of two oblique rafters and a tie-beam; the weight upon the entire truss is transmitted to the walls by the rafters, and the office of the tie-beam is to prevent the lower ends of the rafters from spreading outwards, and thereby overturning the walls or columns sup-

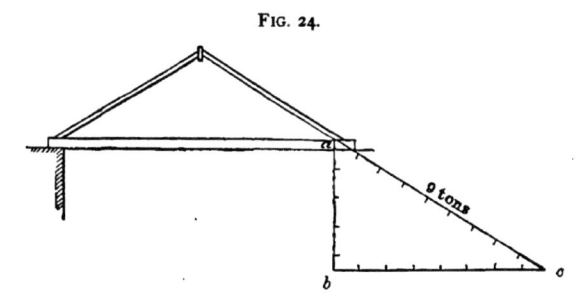

FIG. 24.

porting the truss. Each truss has to support one bay of the roof, and consequently each of the two rafters has to support one-half of the weight of that part of the roof which lies between two adjacent trusses, and as this weight is uniformly distributed, it may be represented as if it really acted at one point; namely, in the middle. The whole structure is balanced or kept in equilibrium, on each side, by three forces: first, the reaction of the wall, which is equal to one-half the weight upon the entire truss; second, the oblique thrust of the rafter; and, third, the horizontal tension of the tie-beam.

It is one of the fundamental principles of mechanics, that if three forces acting upon the same point are in equilibrium, then three lines, drawn parallel to the directions in which the forces act, will form a triangle, the lengths of whose sides are exactly proportional to the magnitudes of the forces. Hence, if three forces act at any point of a braced structure, the directions of which are known, we can determine their relative magnitudes by drawing a triangle. Further, if the magnitude of one of the forces is known, and the side of the triangle corresponding to that force is made equal to it in magnitude, on any scale of equal parts, then the other sides of the triangle will be equal to the other forces, on the same scale.

Referring to Fig. 24, the triangle *a b c* is formed by drawing lines parallel to the directions of the acting forces, one of which passes along the oblique rafter, the other along the tie-beam, the third being the upward vertical reaction of the wall. As the weight upon the wall is the only known quantity, the student will first draw a vertical line *a b* to represent, on any convenient scale, the upward reaction; that is to say, if the weight or upward reaction is $4\frac{1}{2}$ tons, then the length of the line drawn must be made $4\frac{1}{2}$ inches, or $4\frac{1}{2}$ half-inches, or $4\frac{1}{2}$ units of any other convenient dimension. From the ends of the line so obtained, let two other lines *b c* and *a c* be drawn, parallel to the other two forces, so that the three lines may form a triangle; then the length of the sides of the triangle, when measured by the scale used for the vertical force, will represent the exact amount of the thrust upon the rafter and the tension on the tie-beam.

The strains that come upon the diagonals of lattice girders and the more complicated forms of roofs may also be found in a similar manner, but the student is referred to the literature specially devoted to those structures for further information. He is, also, advised to make himself master of the foregoing simple illustration of the truss, because, when it is thoroughly understood, the remainder of the chapter will be found comparatively easy.

Fig. 25 is a skeleton diagram showing the form of trussed beam commonly used for travelling cranes; the stress upon the top and bottom bars of such a structure may be ascertained

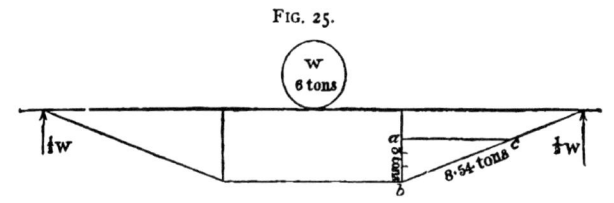

FIG. 25.

in the same manner as for an ordinary flanged beam, and the stress upon the tie-bar may be found by drawing the triangle $a\,b\,c$, as explained in the previous example. Let the beam be 20 feet span and required to carry 6 tons at the centre, with the depth of the truss as usually made, namely, $\frac{1}{8}$ of the span; in this example it is shown to be 2 feet 6 inches.

The stress upon the top and bottom bars will therefore be $3 \times 10 \div 2\frac{1}{2} = 12$ tons, the stress upon each strut is equal to the total weight divided by the number of struts, $6 \div 2 = 3$ tons, and the stress upon the tie-bars is shown by the diagram to be 8·54 tons.

If the load is required to travel from one end of the truss to the other end, as in a travelling crane, then the truss must be counterbraced as shown in Fig. 26, because, when the weight is over the strut B, the tie-rod $d\,e\,$D has a tendency to straighten itself, and thus to force the other strut C upwards, and along with it the top bar. In very light cranes this tendency is usually counteracted by putting sufficient material into the top bar, in order to make it rigid enough to bear the stress without change of form, but in all properly constructed trusses, which are required to support a moving load, counterbraces are added, because by their assistance

216   *On the Strength of Structures.*

the transverse stress upon the top bar is converted into a stress acting in the direction of the length of the component parts of the structure, consequently less material is required

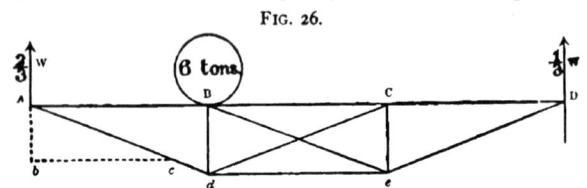

FIG. 26.

in the truss to support a given load, and the tendency to a change of form in the structure is at the same time reduced to minimum.

The reactions at the support will be 4 and 2 tons respectively, and the strain upon the top bar $\frac{4 \times 80}{30} = 10\frac{2}{3}$ tons; the strains upon the strut and tie-bar may be ascertained by the triangle of forces as previously explained, and in the manner shown by the triangle A *b c* in Fig. 26, which determines the strain upon the top bar, strut, and tie-bar, as caused by the transmission of the weight to the support A, the line A *b* being made equal by scale to the reaction of the support A.

The hydraulic wharf crane (Fig. 27) is another familiar example of the employment of a braced structure, the framing by which the top cap and slew drum are carried, forming one side of the triangle, and the tension rods and jib the other two sides. At the jib head three forces meet and are in equilibrium—namely, the downward pull of the weight, the resistance of the tension rods, and the oblique thrust of the jib—and they may be represented by the sides of the triangle *a b c*, Fig. 28, which are drawn parallel to them. If the side *a b* (which is drawn parallel to the downward pull of the weight) be made by scale to represent the weight, the lengths of the other sides *a c* and *b c* will represent the tension upon the tie-rod and the compression upon the jib. In this case, with a weight of 6 tons, the tension is 18 tons, and the compression is $21\frac{7}{8}$ tons, by scale.

On Roof Trusses and Cranes.

To be strictly accurate it is also necessary to take into account the effect due to the tension upon the chain; that is to say, to the vertical stress upon the chain, which is due

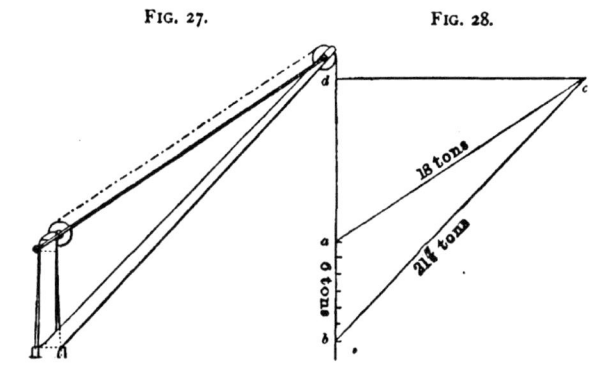

FIG. 27.   FIG. 28.

to the weight suspended, as well as the tension of the inclined portion of the chain, because both unite in their effect upon the sheave-pin at the head of the jib, and thus affect the crane structure in proportionate degree by reducing the tensional stress upon the tie-rods, and by increasing the compressive stress upon the jib structure. The amount of these strains may be ascertained by first finding the stress upon the pin at the jib head, and then resolving it into two component stresses, the one acting in the direction of the tie-rods, and the other in the direction of the jib.

The crane post is subjected to a transverse strain caused by the oblique pull of the tension rods at its upper end; the amount of this pull in the horizontal direction can be ascertained, by resolving the oblique pull into its two component parts, one horizontal or at right angles to the post, and the other vertical. This latter may be disregarded for the present, as it causes no transverse strain upon the post. This resolution may be made as shown on the upper part of

Fig. 28, in which ac represents the pull of the tension rods in direction and magnitude, dc the horizontal component of the pull of the tension rods, and da the vertical component.

The post of this crane may be considered as similarly circumstanced to a cylindrical or other beam, projecting from a wall and loaded with a weight at its outer extremity, equal in amount to the horizontal component of the oblique pull of the tension rods, as found by the diagram. It is also subjected to a compressive stress in the direction of its length.

The compressive stress in the direction of the length of the crane post involves a more recondite calculation; it will be found equal in amount to the weight of the crane structure, plus the effect of the weight lifted; which latter will be equal to the algebraical sum of the vertical components of the oblique pull of the tension rods and the thrust of the jib; that is to say, the vertical component of the former will act upwards upon the post, but that of the latter downwards, and the resulting effect of these two opposite forces will be the remaining portion of the greater stress after the lesser stress has been subtracted from it, because the greater neutralises the lesser, and the remainder only acts upon the post. This force is comparatively so small that the calculation may be disregarded, because if the crane post is strong enough to resist the transverse strain, then the compressive strain will not be called into such exercise as will affect the stability of the structure.

On the Strength of Sheer-legs.

The sheer-legs, or poles, shown by skeleton diagram in Fig. 29, are an example of a temporary braced structure. In this figure B C represents the position of the sheer-legs, and A C that of the tension rod. The tension rod in this case is termed a 'guy,' and may be either a hempen or wire rope, or chain. The holdfast for this guy is furnished by strong pickets driven into the ground, and the thrust upon the poles is supported by a small foot plate, laid upon a platform of timbers at B.

On Roof Trusses and Cranes. 219

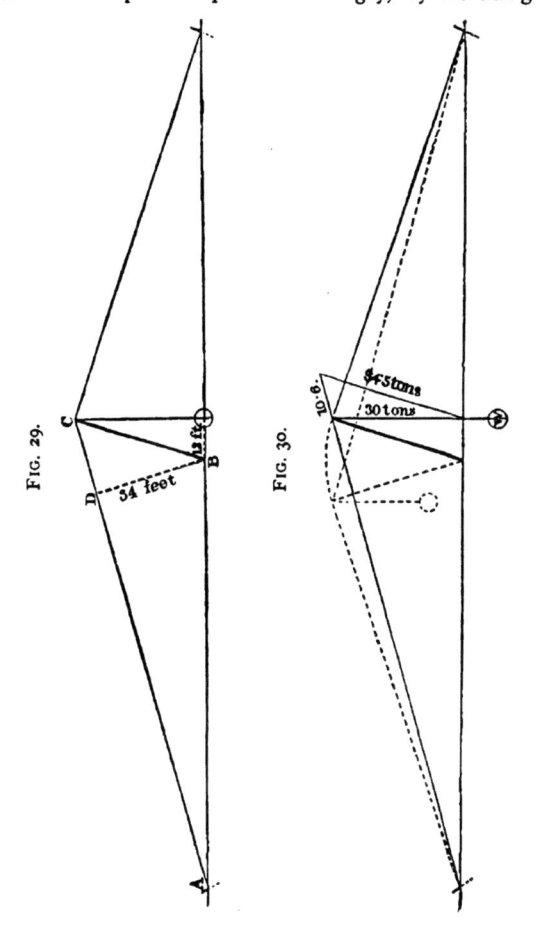

The diagram (Fig. 30) shows the method of ascertaining the strains upon the poles and the guy, by the triangle of forces. With a 30-tons load, hanging 12 feet beyond the perpendicular, the thrust on the sheer-legs will be found to be 34·5 tons, and the tension on the guy 10·6 tons.

The amount of the strain upon the guy may, however, be found by the principle of the lever, or, as it is sometimes called, the principle of moments, as shown by Fig. 29. The pole B C may be regarded as a lever turning upon B as a centre; then the weight multiplied by the perpendicular distance at which it acts from B—in this case 12 feet—and divided by the perpendicular distance of the guy from B, will give the tensile stress upon the guy. The best method to find the length of the perpendicular from the fulcrum B to the guy is to make a skeleton diagram similar to Fig. 29, but to a scale of not less than 5 feet to 1 inch.

The sheer-poles are 40 feet in length, the distance from the holdfast to the bottom of the poles is 120 feet, and the top of the poles is 12 feet beyond the perpendicular line raised from the centre of the feet of the poles; with these data, the length of the perpendicular B D will be found to be 34 feet. Then 30 tons × 12 feet ÷ 34 feet = 10·59 tons, the strain on the guy rope A C. Although there are two guy ropes used with these sheer-poles, there is of course no strain upon the fore guy when the weight is in the position shown on the diagrams.

The two guys are provided to enable the sheers to pick up a weight at the point shown on the diagram, and transfer it to the position shown by dotted lines, the strain upon the one guy rope gradually decreasing till the poles reach the perpendicular position, at which point there is no strain upon either guy; but immediately that point is passed the other guy is subjected to a strain gradually increasing, until, when a point is reached 12 feet upon the other side of the perpendicular, it is of course strained to the same amount as the other guy was in the first position of the load.

To permit this movement and to render it a perfectly safe operation, and also to enable a light hoisting crab, with its small complement of men, to perform the work, a portion of the guy is composed of an ordinary tackle, consisting of two blocks with three pulleys in each, and the necessary length of rope to allow the guy to be lengthened the required amount; the running end of the tackle is wound direct upon the crab barrel. Then as the strain is distributed to six reduplications of the rope, the strain upon each of them and on the rope leading to the barrel will be only $\frac{1}{6}$th of the total strain upon the guy, or 10·6 ÷ 6 = 1·76 ton.

To enable the weight to pass between the poles, they are set 20 feet apart at the bottom; and if the hold fast ends of the guys are fixed on a line bisecting at right angles the line joining the feet of the poles, there will be no lateral strain upon the structure, except such as may be due to an external circumstance, such as a side wind, and which the spread of the feet of the poles enables them to resist effectually.

On the Strength of a 30-Ton Steam Crane.

Fig. 31 shows an arrangement of steam crane, for lifting weights of 30 tons, and as there are in the various parts of this structure examples of transverse, tensile, compressive, and shearing strains, it is selected as an illustration, to show the student a method of ascertaining the amount of the stresses upon such a structure, and also the quantity of material required to resist them.

This crane works upon a centre pivot at B, and is prevented from rising by nuts, which act as a collar, and are fixed to the pivot in such a manner as to prevent unscrewing.

The jib of the crane is upon one side, having two rollers which travel upon a circular rail at A, and is partly counterbalanced by the weight of the steam boiler, steam engines, and platform, but still leaving a weight of about 2 tons, so that the weight of the jib may be disregarded when the crane is unloaded. The steam power is used for lifting or lowering, and is also employed for turning the crane round upon its pivot, by giving a slow motion to one of the rollers, and the apparatus is so arranged that the motion for raising

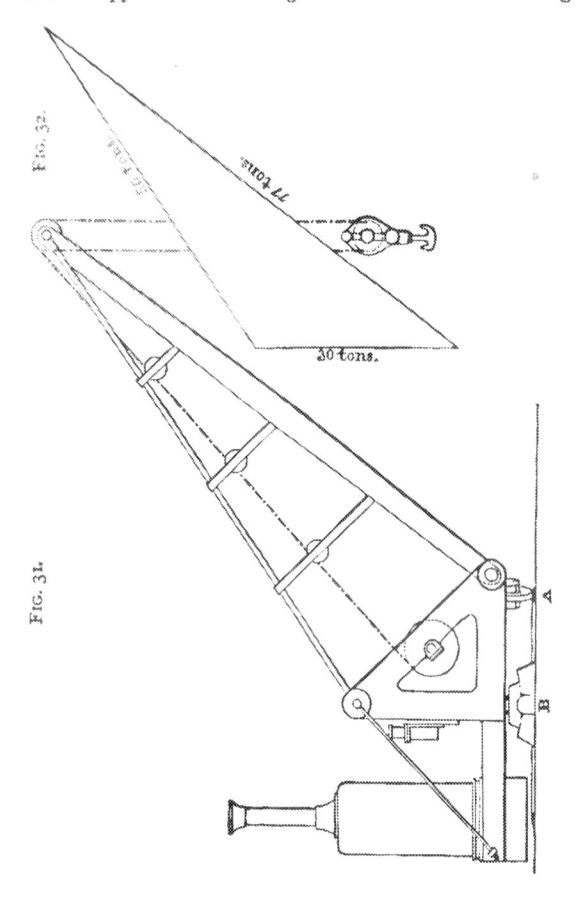

or lowering, and that for turning in either direction, may both be in gear at the same time.

The whole crane structure is kept in equilibrium by three forces—namely, the downward pull of the weight, the resistance to the pressure upon the guide rail A, and the resistance to the upward pull upon the centre pin or pivot B. It may be treated as a large bent lever loaded at the end of one arm with a weight of 30 tons.

The direction of these three pressures are shown in Fig. 33. The weight, multiplied by its perpendicular distance from the guide rail or fulcrum of the lever, and

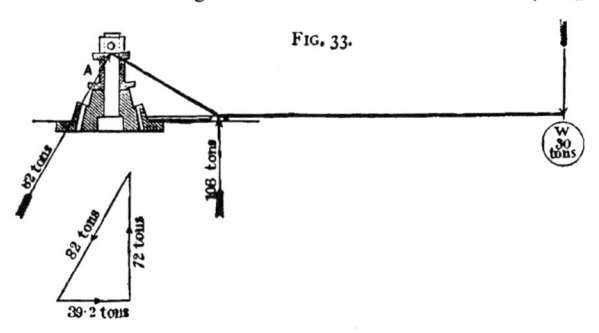

FIG. 33.

FIG. 34.

divided by the length of the other arm of the lever, gives the amount of force that is acting at the short end of the lever, to balance the weight of 30 tons; thus, 30 × 25 ÷ 9·17 = 82 tons nearly, and this force of 82 tons, resolved (as shown by the diagram, Fig. 34) into its horizontal and vertical components, gives 39·2 tons for the former and 72 tons for the latter. To find the pressure upon the guide rail, multiply the weight by its distance from the centre pivot, and divide by the distance from the other end of the lever to the guide rail; thus, 30 × 33 ÷ 9·17 = 108 tons.

224   *On the Strength of Structures.*

Having found the strain upon the centre pin and guide rail, and agreed upon the nature of the foundation best adapted for the situation, it becomes a simple matter to determine the amount of material necessary to resist those strains.

The strains upon the jib and tension rods of the crane may be found by means of a diagram, in a similar manner to that described for finding the strains upon the same parts of the hydraulic wharf crane, and by reference to Fig. 32 it will be seen that they amount to 77 tons compression on the jib, and 56 tons upon the tension rods, while lifting a weight of 30 tons, and the quantity of material required to resist the strains upon the tension rods, when the iron is subjected to a stress of 2 tons per square inch, will be $56 \div 2 = 28$ square inches, and as there are two rods, there will be 14 square inches required in the section of each, and if the rods are round they will be $4\frac{1}{4}$ inches in diameter.

The jib of these cranes is a most important part of the structure, and the great compressive strength of cast iron would seem to indicate that it would be the best material of which to construct it; but there are many difficulties in making a cast-iron jib of such proportions, capable of withstanding with safety the shocks to which all crane structures are more or less subjected, in working, and hence wrought iron is used as it is more reliable.

Long pillars of wrought iron have to be carefully braced to resist the tendency to flexure, and more generally is this the case when they are not perpendicular to the line of thrust, or when they are (as it is sometimes termed) 'bowed.'

Fig. 36 shows an arrangement of wrought-iron pillar very commonly used for the jib of large cranes, and also for the piers of viaducts and other similar structures; it consists of two side plates of sufficient area to resist the compression to which they are subjected, when well braced with lattice bars, dividing the long unsupported column into a number of short pillars, and keeping the side plates in a direct line;

On Roof Trusses and Cranes. 225

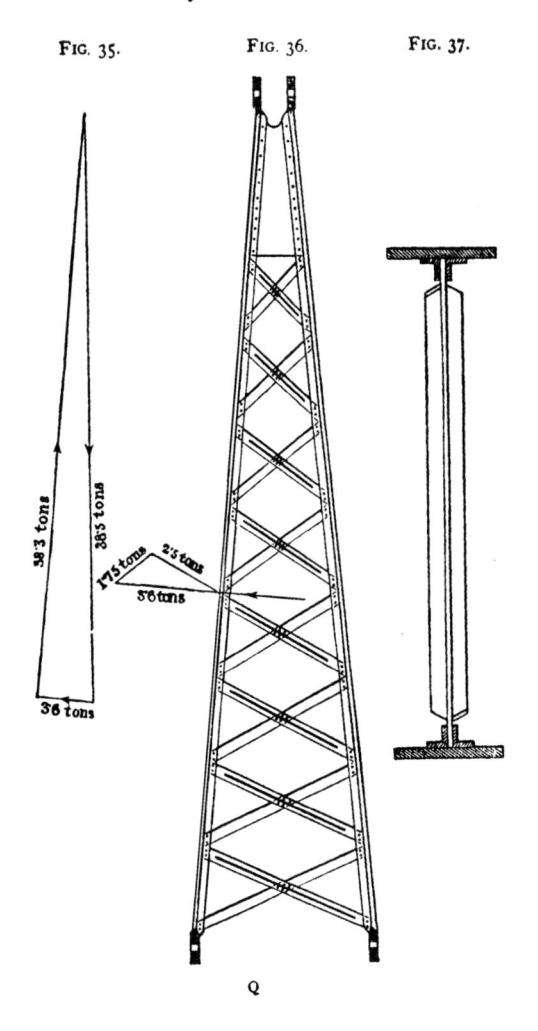

FIG. 35. FIG. 36. FIG. 37.

two angle irons, extending the whole length of these plates, are riveted to their inner side, for the attachment of the lattice bars.

The side plates of this jib are subjected to a transverse strain, along their whole length, tending to produce flexure, the value of which may be found by resolving half the direct pressure on the jib (because there are two side plates) into its two component parts, as shown on the diagram (Fig. 35), the one, acting along the line of the side plate, equal to 38·3 tons, and the other, at right angles to it, which is equal to a compressive stress of 3·6 tons. This latter is the force which tends to produce flexure, and it is to resist this strain that the lattice bars are inserted; these bars are, as shown in Fig. 37, of a **T** section, 6 inches × 3 inches × ½ an inch, and have an effective area to resist tensile strains of 3½ square inches, after deducting the rivet holes at their end.

The tensile strain upon these bars is also shown on the diagram (Fig. 36), and is found by resolving the transverse stress of 3·6 tons upon the side plates into two component forces acting in the direction of the lattice bars, and equal to 2·5 tons on one and 1·75 ton on the other. A simple bar, 3½ inches × ½ an inch, would have been sufficient to resist this strain, and therefore there must be some other reason for making them stronger and of a **T** section, and it is to enable them to prevent the deflection of one bar without the other, and thus to form the whole into a structure capable of resisting the compressive and transverse strains above described, and also the tendency to deflection, or 'sagging,' as it is sometimes called, engendered by its own weight.

The jib and tension rods are connected at the jib head by a strong turned pin, which also carries the two chain pulleys; this pin is subjected to a shearing strain at each end, equal to the pull upon one tension rod, and it should at least be of the same sectional area as the tension rod, and hence of the same diameter; the lower end of the jib is connected to the crane framing by two strong turned pins, fitting into the framing and the side plates of the jib, which are made thicker at this point, and also at the point where the pin at the jib head passes through them, and bored out to the exact size of the pin. These pins are very important parts of the structure, and should be well fitted, so that they cannot bend in the holes through which they pass.

Each pin at the lower end of the jib has to resist a pressure of 38·5 tons—that is, one-half the total thrust—and, in order that the material may not be subjected to a greater stress than 2 tons per square inch, must have a shearing section of $38·5 \div 2 = 19·25$ square inches. But as the pin would have to be cut across at each side of the side plate, it need only have an actual sectional area of $19·25 \div 2 = 9·625$, which is equivalent to a diameter of $3\frac{1}{2}$ inches; but if the pin is not supported at each side, it must then have an area of 19·25 square inches, as above stated, or a diameter of about 5 inches.

Each tension rod is also connected by a turned pin to the top of the side frame of the crane, similarly to the lower end of the jib; these pins are of course subjected to the same stress as the pin at the jib head, but, as they are supported at each side of the eye of the rod, they only require to have half the sectional area; that is, $14 \div 2 = 7$ square inches, or a diameter of 3 inches.

The centre pin, or pivot, upon which the crane revolves, is fitted at the top end with a strong nut, shown on Fig. 33, which is screwed down to its bearing upon the girder in which the pivot works, and may be secured by a steel pin, passing through the pivot and nut, or otherwise, so that the latter cannot work back; and the girder is bored truly and the pivot is turned to fit nicely, so that there may be no play. This pivot and nut have to resist the whole of the upward pull due to the 30-tons weight; the magnitude and direction of this force and its vertical and horizontal components are shown on Fig. 34. The latter of these strains acts with a leverage equal to the whole length of the pivot, tending to break it across in a similar manner to that described in the example of the wharf-crane post ; and if the pivot is made strong enough to resist this stress, it will be of ample sectional area to resist the vertical stress, which causes only a tensile strain upon the material.

The formulæ generally given, to find the diameter of round bars of iron required to carry a weight in this manner, are founded upon two facts : first, that a bar of wrought iron 1 inch in diameter and 12 inches long, supported at each end and loaded at the centre, will break with a weight of about 2,000 lbs. ; secondly, that the strength of such bars varies directly as the cube of their diameter and inversely as their length.

We know that a bar or beam, similar in every respect to the above, but supported at one end and loaded at the other, will only support $\frac{1}{4}$th of that weight, or 500 lbs. ; and if a 1-inch bar, 12 inches long, breaks with 500 lbs., a similar bar, 32 inches long, (the length of the centre pin) will only bear $500 \times 12 \div 32 = 187\frac{1}{2}$ lbs. Therefore, the centre pin would require to be equal in diameter to the cube root of its breaking weight, divided by $187\frac{1}{2}$, the breaking weight of a 1-inch bar, or, which is the same thing put in another way, the cube of the diameter of the centre pin, must be equal to the breaking weight in lbs. which it will have to support, divided by $187\frac{1}{2}$.

The breaking weight should be 392 tons—that is, ten times the working stress of 39·2 tons, and $392 \times 2240 \div 187\frac{1}{2}$ = 4683—and the cube root of 4683 = 16·73—say, $16\frac{3}{4}$ inches—the diameter of pivot required, if it did not fit the hole in the girder.

By making the pivot to fit the hole in the girder exactly, the load upon the pivot would be distributed over its entire length, and its strength would thereby be doubled, and hence the cube of the diameter of the pivot required would be only one-half of that required under the former circumstances.

On Roof Trusses and Cranes.

The cube of the diameter in the former case is 4683, and in the latter case would be 4683 divided by 2, which is equal to 2341·5, and the cube root of 2341·5 is 13·26— say, 13¼ inches—the diameter of pivot required, provided it is made to fit in the girder exactly.

The sectional area of pivot required to resist the vertical component of the load only, supposing the material to be subjected to a stress of 2 tons per square inch, would be 72 tons ÷ 2 = 36 square inches, or an equivalent diameter of 6·8 inches nearly; so that if sufficient material is provided to resist the lateral or transverse stress, there will be ample to resist the vertical or tensile stress.

The cast-iron girder A, Fig. 33, in which the pivot is fitted, is 8 feet long between the supports, and 2 feet 6 inches deep, and has to resist the same strains as the pivot—that is, a pressure at the centre of 72 tons, the vertical component of the upward pull of the crane, and, as previously explained, in the case of the common cast-iron girder at p. 185, the bottom flange will, by the principle of the lever, be subjected to a stress caused by half the weight upon the girder, multiplied by half the span, and divided by the depth of the girder—thus, 36 × 4 ÷ 2½ = 56·4 tons, and allowing that cast iron in a crane structure should not be subjected to a tensile stress of more than 1 ton per square inch of section, the flange must contain 56·4 square inches of iron in its section : each flange has to resist the same amount of lateral strain, and it would therefore be advisable, in this case, to make them of the same sectional area at the centre, and reduce them each to ⅔rds the area at the ends.

The girder above the guide rail is cast hollow, so that the upper half of the guide wheels may work within it; the transverse section is shown in Fig. 39, and the side elevation in Fig. 41. This girder has to resist an upward force equal to the pressure upon the guide rail, shown by the diagram (Fig. 33) to be equal to 108 tons; the top flange of this girder will be subjected to a tensile strain, because the load upon it acts vertically upwards, instead of downwards, as in the case of any ordinary weight.

This girder is also 8 feet span and 2 feet deep, and it is loaded at two points; namely, directly above each guide wheel at 2 feet from the ends, with a pressure at each point of half the total load, equal to 54 tons, as shown in Fig. 40. Each of these forces is equal in its effect upon the girder to 27 tons, acting at the centre of its length, because, although the weight is reduced one-half, its distance from the support, and consequently its leverage, is doubled, and a total force of twice 27 = 54 tons, acting at the centre of the girder, would be precisely equivalent to the two forces of 54 tons acting as shown in the diagram. By a similar process to that pursued in finding the strain upon the flange of the girder A, we find, that this weight of 54 tons would cause a tensile strain of 54 tons upon the top flange of this girder—that is to say, the weight upon one support, due to the 54 tons acting at the centre of the girder—namely, 27 tons, multiplied by half the length of the girder, 4 feet, and divided by the depth, 2 feet, = 54 tons—and therefore the sectional area required in the flange to resist it, when the iron is subjected to a stress of 1 ton per square inch = 54 square inches. As there is no lateral strain upon this girder, the bottom flange may be reduced to $\frac{1}{4}$th the area of the top, or $13\frac{1}{2}$ square inches; but as one-half of this flange will be placed on each side of the guide wheel, the webs should be connected by a cast-iron rib across the middle of the girder, as shown by the dotted lines in Fig. 40.

The side frames of a crane are necessarily made of such a form as may be best adapted to give room for the various bearings and supports for the connection of the jib, tension bars, and connecting girders. The principal stress, upon the side frames of these cranes, is that which is caused by the pull of the tension rods at the top, and these frames may be considered to be subjected to a similar strain to a wall crane, when lifting a weight at a distance beyond its proper radius, as shown

Crane Gearing. 231

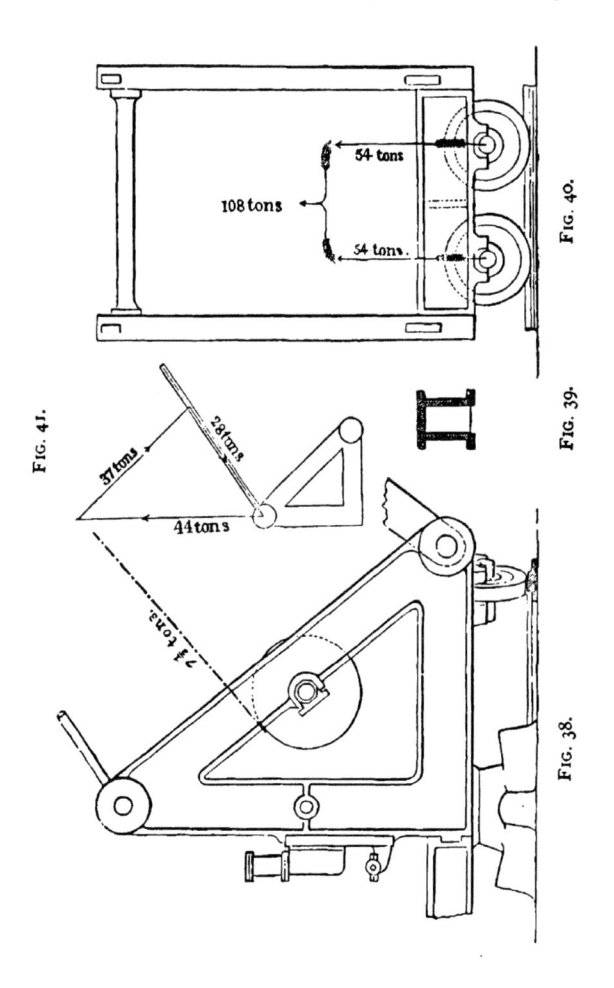

in Fig. 41, the horizontal members of these side frames being in a similar position to the wall upon which the crane is fixed, and the other members fulfilling the same office as the tension rod and strut of the wall crane, the pull of the tension rods of the large crane representing the weight.

The total pull exerted by the tension rods is 56 tons ; that is, 28 tons at the top of each of the side frames. This force is balanced by the resistance to tension supplied by the vertical member, and the compression of the oblique member, and if, as before explained, lines are drawn parallel to the direction of the three forces, and the one representing the pull of the tension rod be made equal by scale to 28 tons, the length of the other lines, measured by the same scale, will represent the strains on the vertical and oblique members, as in Fig. 41. By the diagram, these forces are found to be 44 tons on the vertical, and 37 tons upon the oblique member of the frame, and hence an effective area, when all holes are deducted, of 44 square inches, will be required to meet this strain alone in the vertical member, and additional material to resist the load caused by the gearing shafts.

The oblique member has, in addition to the compressive strain caused by the pull of the tension rods, to withstand the pull, at the centre of its length, of the hoisting chain upon the barrel, which amounts to $7\frac{1}{2}$ tons, as shown in Fig. 38.

The shafts which carry the gearing of a heavy crane are of great importance, and the method of finding the diameter of shaft, required to resist the strains, can be best shown by a few examples; we may take, for instance, the barrel shaft, and the shaft of one of the guide wheels, which is fitted with gear and driven to slew the crane round. The principal strain to which these shafts are subjected is that of torsion, and the torsional strength of cylindrical shafts varies as the cube of their diameters; this knowledge, coupled with the fact that a bar of wrought iron 1 inch in diameter is twisted asunder by a weight of 800 lbs., acting at the end of a

12-inch lever fixed upon one end, enables us to determine the size of shaft required to resist any torsional strain.

The barrel shaft of this crane has to resist a twisting force of 7½ tons, acting upon a barrel 4 feet diameter, the length of the lever being one-half the diameter or 2 feet, and the breaking strength of the shaft should be equal to ten times that amount, or 75 tons acting with a 2-feet leverage. The breaking strength of a 1-inch bar with a 2-feet lever will be 800 ÷ 2 = 400 lbs., and hence by dividing 75 tons reduced to pounds by 400, we obtain the cube of the diameter of shaft required. Thus 75 × 2,240 ÷ 400 = 420, and the cube root of 420 = 7·488, say 7½ inches, the required size of shaft.

The load upon the guide-wheel shaft is determined by the adhesion due to the weight upon the wheel, which under favourable circumstances amounts to 600 lbs. per ton of the load, the weight upon the wheel is 54 tons, and 54 × 600 = 32,400 lbs. as the adhesion of the wheel upon the guide-rail. The diameter of the wheel is 3 feet, and therefore the length of lever 18 inches; the weight required to break a 1-inch bar with 18-inch lever = 800 × 12 ÷ 18 = 533 lbs. The breaking strength of the shaft should be ten times the adhesion acting at the end of an 18-inch lever, that is 32,400 × 10 = 324,000 lbs., and 324,000 ÷ 533 = 608, the cube of the diameter of shaft, and the cube root of 608 = 8·47 inches, the diameter of shaft required.

The size of the remaining shafts is determined in the same manner, that is, by first finding the twisting load, then the strength of a 1-inch bar under the same conditions, dividing the former by the latter and extracting the cube root of the quotient.

Referring to the remarks made at page 191, on the comparative strength of shafts with overhung gear, and to gear having an outer bearing, let us apply them in such a crane as that now under consideration.

Let us suppose the pinion which gears into the wheel

on the barrel shaft, namely, the main driving pinion of such a crane, to be keyed upon the outer end of a shaft, say 4 inches in diameter, and that the pinion is 12 inches wide and has to drive the large spur wheel, with a load of $7\frac{1}{2}$ tons at the pitch line. We have already said, that a bar of wrought iron 1 inch in diameter and 1 foot long, supported at both ends, will be crippled with a weight of 2,000 lbs. acting at the centre. Now a similar bar, supported at one end only, and having the load distributed as in the case of the above shaft, would only support one-half of 2,000 lbs., or 1,000 lbs., and as the strength of cylindrical beams varies as the cube of their diameters, all other conditions being the same, the above-mentioned shaft would be crippled with a weight of 1,000 lbs. multiplied by the cube of its diameter, or 1,000 × 64 = 64,000 lbs. The actual load on the shaft is $7\frac{1}{2}$ tons, or 16,800 lbs., and hence the ratio of the working to the breaking stress is, as 16,800 is to 64,000, or as 1 is to 3·8.

Such a margin of safety, although sufficient for many purposes, is not so in the case of a crane. The breaking strength (as has been several times stated) should be at least ten times the working load, to ensure perfect safety.

This margin of safety is more especially required when the load is being lowered, under the control of the friction break, and then suddenly checked; the effect of the load, in consequence of the stoppage of the motion, is increased to an extent which may imperil the whole fabric.

This margin would be more than obtained, by simply increasing the length of the 4-inch shaft, and putting a bearing outside the pinion instead of having it overhung, as the breaking stress of a 1-inch bar, under these altered circumstances—that is to say, when supported at both ends and with the load distributed—would be 2,000 × 2 = 4,000 lbs., and hence the strength of the given shaft would be 4,000 × 64 = 256,000 lbs., and the ratio of the breaking strength to the working-load would be as 256,000 is to 16,800, or as 15 is to 1.

Crane Gearing.

Although, when the pinion is overhung, the arrangement may not necessarily cause the breaking of the shaft directly, still it will permit it to deflect and thereby cause the load to be thrown upon one end of the tooth, namely, upon that which is next to the bearing, instead of upon the whole length as it ought to be, and will thus endanger the safety of the gearing; and even if this latter should be of ample strength, the deflection of the shaft will continue, until in time the shaft will take a permanent set, and ultimately break off, close to the bearing.

The method of finding the strength of the teeth required in the respective wheels has been explained in Chapter XIII., p. 193.

Having thus far shown how to find the quantity of material required in the various parts of the crane, we have now only to deal with the chain for carrying the weight, and by general consent the Admiralty formula given at p. 159 is adopted.

In these cranes the weight or load upon the chain barrel is diminished, by multiplying the number of chains to carry the required weight. The total weight of 30 tons is supported by four chains, thus: The chain is anchored to the underside of the jib-head, and passes down under one of the pulleys in the movable block, up to and over a pulley at the jib-head, and again descends, passes under the second pulley in the movable block, and finally ascends and passes over a second pulley at the jib-head and thence to the barrel.

The load on each of the lengths of chain above the movable block will, therefore, be 30 tons $\div$ 4 = $7\frac{1}{2}$ tons.

Another way of finding the strain on the chain, which may be used when there is any exceptional arrangement of the blocks or tackle, is to see how much chain is taken up in lifting the weight 1 foot; then if, as in this example, 4 feet of chain are taken up on the barrel, the load upon the chain will be $\frac{1}{4}$th of the total weight lifted; if 6 feet be taken up, the load will be $\frac{1}{6}$th, and so on; for, by a well-known principle, a small weight acting through a long distance is precisely equal to a large weight acting through a proportionately shorter distance.

On Crane Foundations.

In order that such cranes may be reliable, it is necessary to have a sound and unyielding foundation. Where the ground is firm and solid, there is comparatively little difficulty in obtaining the required stability, but, as a rule, in the places in which cranes have to be erected, it is otherwise. To meet the requirements fully, the foundation should be such, that the pressure on the guide wheels does not cause the guide rail to yield perceptibly, as it passes round the circle; a yielding rail adds considerably to the motive power required to turn the crane round.

In Figs. 42 to 49 are shown three descriptions of foundation, which have been constructed for 30-ton cranes. The foundation illustrated in Figs. 42 and 43 is adapted for a crane situated upon an ordinary solid wharf; that in Figs. 44 and 45 is for a crane which had to be erected at some distance from a wharf, in order to obtain a sufficient depth of water to allow vessels of the larger class to come under the crane. From the isolated position of this crane, sufficient mass had to be placed in the foundation to render it perfectly stable and capable of resisting any strain tending to overturn it, without throwing any strain upon the light jetty, erected to carry the loads to and from the crane. The foundation shown in Figs. 46, 47, 48, and 49 is for a crane erected at the head of a long pier.

In each of these three descriptions of foundation, sufficient weight or resistance is provided in order to counterbalance the upward pull of the centre pin, on which the crane turns, and also for any increased strain that might be caused by the sudden stoppage of the descending weight, which is the chief cause of accidents. Hence, although the actual strain

Plan

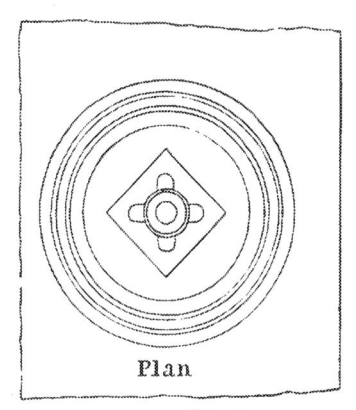

FIG. 42.

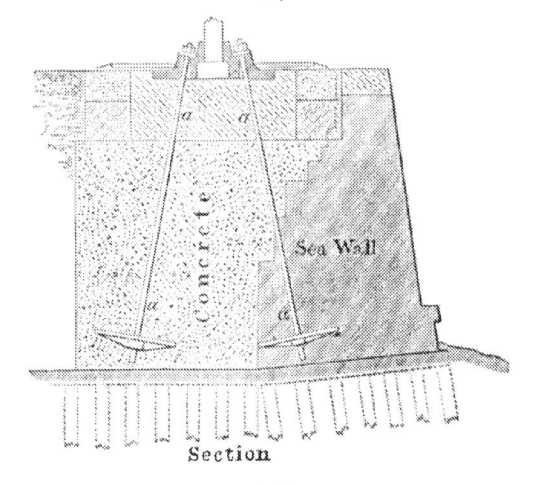

Section

FIG. 43.

is 82 tons, the weight or resistance of material which would have to be lifted, before the crane could overturn, is 328 tons, or four times the amount of the ordinary working strain.

Concrete is used in two of the foundations (Figs. 43 and 45) to furnish the necessary weight, and the cast-iron block, in which the centre post is fixed, is secured to the mass by means of four holding-down bolts *a a*, each bolt being fixed to a strong cast-iron plate 6 feet square, which is embedded in the bottom of the concrete mass, and drawn up tight by strong nuts fitted upon the upper end of the bolts.

In each case, these holding-down bolts have to resist an ordinary working load of 82 tons (Fig. 33), and as they may, by the carelessness of the breaksman, have to resist a much greater strain, they should not in ordinary working be called upon to bear more than 2 tons per square inch of their sectional area, when each bolt will have to resist a vertical strain of 72 tons $\div 4 = 18$ tons, as will be seen by Fig. 34. This vertical strain, in consequence of the oblique position of the bolt, produces on it a tensile strain of 19 tons, and therefore they must each have a sectional area of $19 \div 2 = 9\frac{1}{2}$ square inches (the stress per square inch to which the iron is to be subjected being 2 tons). To obtain $9\frac{1}{2}$ square inches of section, the diameter must be $3\frac{1}{2}$ inches nearly; the actual diameter of the bolts used was 4 inches.

The second class of foundation (Figs. 44 and 45) depends, entirely, upon its mass for stability. It consists of a cast-iron cylinder 20 feet in diameter, and 36 feet high, filled up to within 5 feet of the top with concrete, the remaining space being occupied by masonry to form a bed for the centre block and guide rail; the total weight of the cylinder, its contents, and the crane, is 690 tons, and a force equal to a weight of 300 tons, hanging from the crane-hook, would be required to balance it; for, by taking moments about the bottom edge of the cylinder, we have on the one side 690 tons multiplied by 10 feet, and upon the other 300 tons multiplied by 23 feet, each of which equals 6,900. Hence

Half Plan

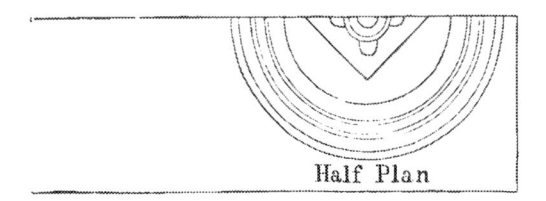

Fig. 44.

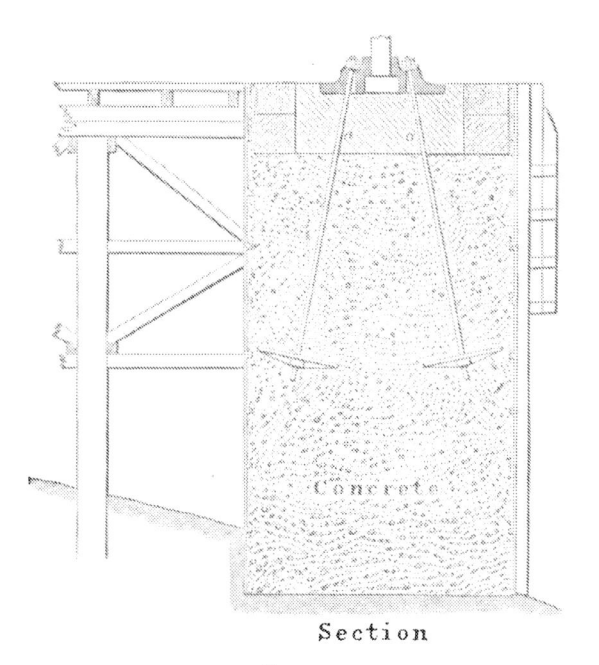

Concrete

Section

Fig. 45.

the whole structure is just balanced; any excess of weight above the 300 tons, upon the crane, would overturn the whole foundation.

Before erecting such a mass, the bottom upon which it is to be built should be carefully examined, and if it is not sufficiently hard to resist the weight, it must be piled, as shown in the wharf foundation, Fig. 43; in the case of the cast-iron cylinder foundation, the bottom was solid rock, and consequently quite able to resist the pressure, which amounted to 2·3 tons per square foot of surface.

In the construction of a large cast-iron cylinder, much judgment is required. A comparatively weak cylinder would suffice, after it is erected and filled in with concrete, but the dangerous period in its history is during the time of its erection, when it is empty and subject to the pressure of a rising tide, tending to crush it inwardly by a collapse of the fabric. To meet this, the joints of the plates ought to be broad and planed, so as to have a uniform bearing, and to fit all over the surface, in order that the arch, formed by the cylinder, may be rigid. The reason of this will be more apparent to the student after reading the next chapter, where reference is made to the collapse of steam-boiler flues, and the laws by which they are governed.

In the third class of foundation, shown in Figs, 46, 47, 48, and 49, which has been erected at a pier-head, the conditions are more complicated, and will require some attention to enable them to be understood.

The centre support consists of a cast-iron cylinder, seven feet in diameter. This cylinder is carried down to the solid gravel, and is filled with concrete. The top of this cylinder is made to receive the lower ends of the four holding-down bolts, *a a*. It will be evident that this centre pillar would be insufficient of itself to furnish the necessary stability, not only in consequence of its want of breadth of base, but likewise on account of the great intensity of the pressure at the base. It will be seen, by referring to Fig. 47, that a

Crane Foundations.

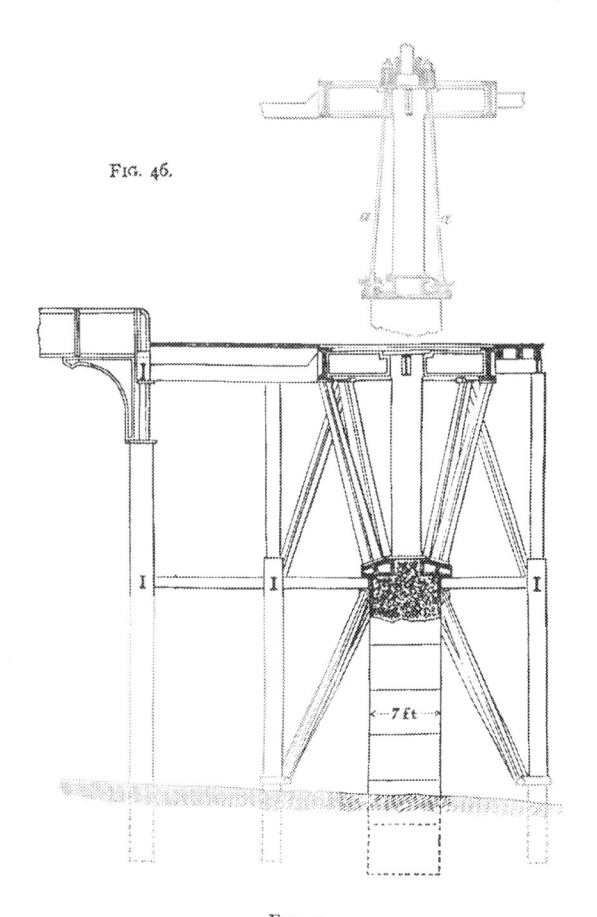

FIG. 46.

FIG. 47.

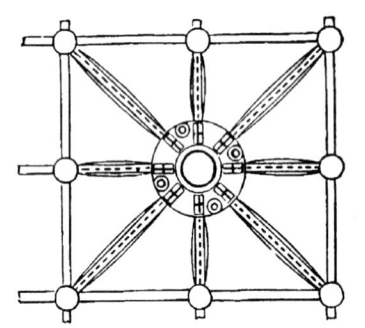

FIG. 48.—Plan on Lower-tier Girders.

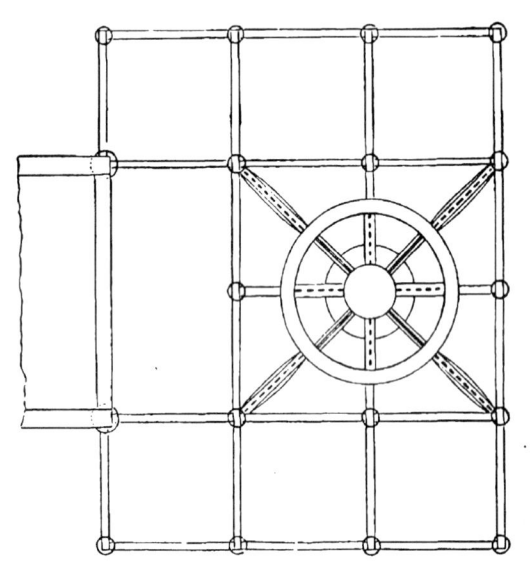

FIG. 49.—Plan on Upper-tier Girders.

Crane Foundations. 243

portion of the weight of the foundation of the pier crane structure is thrown upon the eight surrounding piles, each of which is fitted with a screw blade $3\frac{1}{2}$ feet in diameter. These eight piles have therefore a bearing surface of 80 square feet, and together with the 7-feet cylinder represent a bearing surface of $118\frac{1}{2}$ square feet.

The wrought-iron guide rail, upon which the crane revolves, is laid upon a strong cast-iron ring or girder, and the load is transmitted from it to the central cylinder and the surrounding piles, by twelve cast-iron struts; eight of these struts rest upon the central cylinder, and the remainder upon the lower tier of girders and against the four corner piles of the surrounding square. The strains upon these struts are shown in Figs. 50 and 51. The weight upon the guide rail is 108 tons, distributed by the two guide wheels or rollers, and it is supported by at least two of the central struts in any position of the crane; and when the crane is directly over one of these struts, one half of the load is carried by the two adjacent struts, and the other half by the strut directly under the centre of the crane; hence the greatest load upon either of the struts is equal to $108 \div 2 = 54$ tons of vertical pressure.

These struts are in a similar position to the jib of a crane, and the pressure upon them may be found by a triangle of pressures $a\,b\,c$ (Fig. 50), as explained in the example of the wharf crane. The amount of this strain is shown by the diagram to equal 56 tons, and as the length of the strut is only about eighteen times its least breadth, it may be subjected to a stress of $1\frac{1}{2}$ ton per square inch of its sectional area, at the centre of its length, with safety; and the sectional area required at the centre will be $56 \div 1\frac{1}{2} = 37\frac{1}{3}$ square inches. The actual form of section is shown in the figure, and the actual area is equal to $39\frac{3}{4}$ square inches.

When the crane is directly over the diagonal of the square formed by the surrounding piles, the vertical pressure of 54 tons is supported by one central and one diagonal strut,

14 *On the Strength of Structures.*

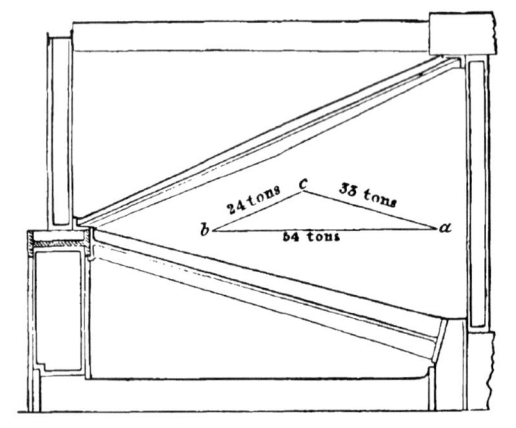

FIG. 50.

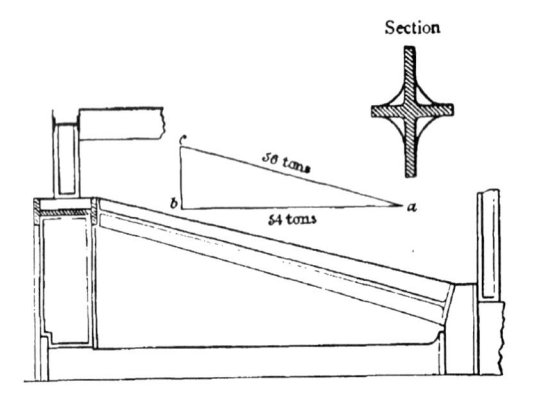

FIG. 51.

Crane Foundations. 245

and the strains upon each are shown, by the diagram $a\ b\ c$, (Fig. 51) to equal 33 tons on the former and 24 tons on the latter.

The weight of the crane and its foundation, including the cast-iron cylinder filled with concrete, is about 190 tons; this, with the weight lifted, is equal to a total of 220 tons, a portion of which is transmitted to the surrounding piles by the eight diagonal struts, already referred to, abutting upon the underside of the lower tier girders, and fixed at their lower extremity to the piles, by a joint pin 3 inches in diameter.

These eight struts have each to support a vertical pressure of $220 \div 8 = 27\frac{1}{2}$ tons; and as they rest upon a joint pin at one end, they are only capable of carrying $\frac{2}{3}$rds the load of a similar strut with both ends fixed, and therefore must not be subjected to a greater stress than 1 ton per square inch of sectional area at the centre. By diagram, as shown in Fig. 53, the longest struts, namely those fixed to the corner piles, are subjected to a compressive stress of $33\frac{3}{4}$ tons by the action of the vertical pressure of $27\frac{1}{2}$ tons upon the upper end, and hence the sectional area required is $33\frac{3}{4}$ square inches. The actual form of section is shown in Fig. 52, and it has an area of 36 square inches.

If the student has been able to understand this chapter, he will be in a position to apply the knowledge acquired in other directions. The elementary principles here explained are not by any means confined to the examples that have been selected for illustration; they are equally applicable to a variety of other structures. The remarks that have been made, upon the direction and amount of stress which is brought to bear upon the members of roofs and cranes, relate also to any other structure in similar circumstances, and it will be useful for the student to apply the knowledge gained in other cases with which he has to deal, so as to fix the principles in his mind. The same may be said of the observations on form and proportions, and on the quantity of material to be introduced so as to afford the requisite

246 On the Strength of Structures.

strength, that the structure may not only be able to balanc
the stress, but to have in addition the necessary margin
safety. The margin of safety necessary in different cases
a subject which admits of great difference of opinion ; in th

FIG. 52.

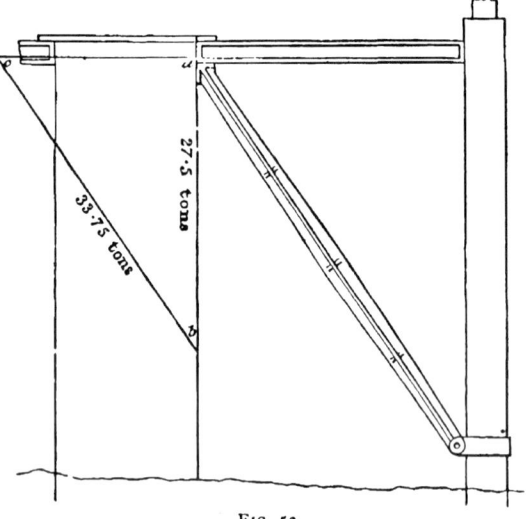

FIG. 53.

volume a leaning to extra caution has been chosen with
purpose, so that if the student errs at the outset of life,]
may err on the safe side ; and as he gains experience, tl
margin of safety can be modified if it is found advisable.

CHAPTER XVI.

STRENGTH OF RIVETED STRUCTURES—STEAM BOILERS, ETC.

BUILT-UP structures, such as steam boilers, which are made of thin plates, riveted together by clenched iron rivets, depend for their strength upon a number of conditions. As a rule, the boiler plates have not the same ultimate tensile strength, per square inch, in the direction of the fibre, as the same iron when made into the form of thick bars. The difference amounts even to several tons, and still more unsatisfactory is the reduction of strength in the other, or transverse direction. This limited tenacity in boiler plates is no doubt due to the rolling of the plates to such thinness, but the weakness must be recognised in estimating the conditions of strength in a steam boiler.

In the chapter on wrought iron, it will be seen that from 23 to 25 tons, per square inch, is the usual tensile strength of such a quality of wrought iron as is required for this purpose; but when this quality of iron is made into thin boiler plates, the tenacity is frequently reduced to 20 or 21 tons, and is seldom over 22 tons; and in the other, or transverse direction, it is rarely over 19 tons per square inch of sectional area.

An important practical lesson to be drawn from this circumstance is, that the boiler-maker, in arranging the order of the plates for the construction of a boiler, should have their strongest direction put in the line of the greatest stress. Take, for example, the shell of an ordinary cylindrical boiler; the least strain is in the direction of the length, hence the strongest direction of the plates should be put circumferentially. This is also the better arrangement, on account of the greater extension or wire-drawing that the iron will submit to, when it is drawn or stretched in the direction of the fibre. For reasons to be shown hereafter, boilers are seldom found to give out in the direction of the length, from constructive weakness; that is, if they are sufficiently strong in the other direction, namely, circumferentially.

When these boiler plates are joined together by the usual butt or lap joints and riveting, the joint becomes still weaker than the plates across the fibres. In thus joining the plates together, by a butt or lap over and rivets, the strength will obviously depend on the resistance both to shearing and to tearing. The rivets may be considered as pins which may be shorn, the force which is required to shear a rivet being the shearing strength of the iron, per square inch, multiplied by the sectional area of the rivet. The tearing of the plates may occur in two ways, either by the plate tearing along the line of rivet holes, or from the pieces between the hole and the edge of the plate being forced outwards, by the simple detrusion of the stronger rivets, the force required being as the shearing strength per square inch, multiplied by the area of the pieces thus pushed out of place.

The shearing strength of rivets depends on the quality of the iron, and ranges from nearly 23 tons, in the very best descriptions, down to 18½ tons; but 22 tons per square inch is usually considered to be about the average strength of the best Yorkshire iron suitable for rivets. The iron of the best plates is about the same in quality as that of the rivets, and, in the direction of the fibre, may be considered to have a tenacity of 22 tons per square inch.

In order to have the rivets equal in strength to the plates, it is necessary to observe a definite proportion between the diameter and pitch of the rivets and the thickness of the plates. The usual rule is to make the diameter of the rivet equal to two thicknesses of the plate, which, as regards strength, is a small fraction under the correct equivalent, but is sufficiently near for all practical purposes.

The plates at the lap or butt being double, will be about equal, in strength, to the rivet, if the space between the hole

Strength of Riveted Structures—Steam Boilers, &c. 249

and the edge is equal to the diameter of the rivet; hence a single riveted joint has a breadth equal to three diameters of the rivet, and the pitch, or distance from centre to centre of rivets, is three diameters. From this it will be seen that one-third of the metal has to be cut out of the plate, and its strength is thereby reduced one-third. This diminution of strength may also be aggravated by punching and rough usage in putting the plates together for riveting; the tenacity of plates, originally 22 tons, may thus be brought down to 18 tons, but this may be avoided by piercing the holes with a drill.

The following Table is taken from the Proceedings of the Mechanical Engineers, for 1872; it forms part of a paper which was read by Mr. Walter R. Browne, bearing on the subject of riveted structures, containing the results of experiments made by Mr. Kirkaldy.

TABLE XLIII.
Riveted Joints.

Description of joint.	Riveting.	Rivet holes.	Diameter to thickness.	Lap or cover to diameter.	Pitch to diameter.	Ratio of strength of joint to that of plate, per cent.
				Lap.		
Lap	Single	Punched	2	3	3	55
		Drilled	2	3	2⅗	62
				Chain. Zig.		
Lap	Double	Punched	2	5½ 6	4½	69
		Drilled	2	5 5½	4	75
				Cover.		
Butt, 1 cover	Single	Punched	2	6	3	55
		Drilled	2	6	2¾	62
Butt, 1 cover	Double	Punched	2	11 12	4½	69
		Drilled	2	10 11	4	75
Butt, 2 covers	Single	Punched	1¼	6	3¼	57
		Drilled	1¼	6	3	67
Butt, 2 covers	Double	Punched	1¼	11 13	5¼	72
		Drilled	1¼	10 12	4¾	79

250   *On the Strength of Structures.*

The resistance to shearing is increased by the frictional force, arising from the grip, due to the contraction of the hot rivet in cooling, and to the action of the hammer or riveting machine. This frictional adhesion is very considerable, but when added to the resistance to shearing it does not make the joint equal in strength to the plate, even in its weakest direction, which should be noted.

From a number of valuable experiments, made by Sir W. Fairbairn, it appeared that, assuming the strength of the plate to be 100, the strength of an equal length of the single riveted lap joint would be 56; but by having a double row of rivets applied, either to form zigzag or chain riveting, the strength was increased to 70. Putting it in another form, and assuming the strength of the iron to be 50,000 lbs. per square inch, then the strength of various kinds of riveted joints would be as under—

	lbs.	
Iron	50,000	
Double riveted joint	35,000	per square inch of the section of the plate.
Single riveted joint	28,000	

The reduction of strength due to riveting greatly affects the construction of boilers. The joint is evidently the weak link of an important chain; consequently the strength of the boiler is only equal to that of the joints; the greater strength of the plates is so much useless metal, which it is the object of the engineer to eliminate, so far as he may be able to do so, by the practical devices which are at his disposal. This may be done either by increasing the strength of the joints, so as to bring them up to that of the plates, or by reducing the substance of the remainder of the plates, or otherwise, the desideratum being uniformity of strength throughout the entire boiler structure.

Since the above facts were made known, a considerable amount of attention has been directed to the subject, in order to devise some arrangement, whereby the strength of

Strength of Riveted Structures—Steam Boilers, &c. 251

the joint might be brought up to that of the plate. With this object in view, it has recently been suggested, to strengthen the joints of boilers and other riveted structures by modifying the shape of the rivet. Rivets are usually round, and about ¾-inch in diameter; if, however, the same quantity of iron was made into an oval form, with the thickness of the narrow part ½ inch, the remainder of the metal going to increase the length in the other direction, the strength of the joint would be altered for the better, because a larger area would be left between the rivet holes.

The resistance to shearing of an oval rivet will be proportional to its area, and will be nearly alike in either direction, as will be seen by referring to the chapter on shearing, p. 149. It is there shown that greater stress was required to shear a bar when placed on its edge than when laid upon its side, but it is more than probable that the increase of resistance was due to the shorter period occupied in the performance of the operation, in the former case, with shears in which the blades were not perfectly parallel. This conclusion is supported by the general result of the experiments made on shearing and punching, which go to show that the stress required is as the area of surface which has to be detruded.

With the boiler plates the case is different, because the weakening of the plates is in proportion to the area punched out, in the line of fracture through the rivet holes; therefore, if a rivet, with the same strength as a ¾-inch cylindrical rivet, can be put into a hole which is only half an inch wide, measured along the joint, it is evident that, so far as this part of the joint is concerned, the plates would be made stronger by one-sixth, all other conditions remaining the same. At the same time the rivet can be strengthened to the same extent, by increasing its substance in the longer dimension, so as to make the plates and rivets equally strong at the point of junction, so far as the resistance to shearing depends on the sectional area. But here another condition must be taken into account; a very narrow rivet acts injuriously on the plate, consequently the limit of this narrowing of the rivet is soon reached, because when it becomes too thin, its wedge action comes into play with greater force, from want of the necessary breadth of bearing surface upon the plate, on the same principle that the diameter of the bolt or pin of a tie rod requires to be of larger sectional area than the rod itself, or that the double rope sling over a crane hook does not afford the double strength of two single ropes, which it always does when passed over a pulley of proper diameter, thereby avoiding the wedge action, due to the concentration of force upon a small bearing surface.

Should such an arrangement of rivets be found to give satisfaction, there will be no practical difficulty in carrying out the system; an elongated hole can be punched as easily as a round hole; nay more, an oval hole can be drilled as easily and as cheaply as a round hole, when proper appliances are provided. This latter statement may be doubted by those who are only familiar with the piercing of holes of the round shape, but such is the case, nevertheless, and as more refined systems of manufacture are introduced, and the forms given to boilers are simplified, the whole row of oval slits may be made at the same time, by a corresponding number of revolving chisels, the plate meanwhile having the requisite amount of transverse motion to give the elongation.

The arrangement of joint-construction is also being modified for the better in other directions. By looking at the section of a joint of the best form, when the ends of one plate or of two plates butt, and are riveted to one cover-plate, it is evident that the rivet is not in the best condition to do its full duty, that it would be stronger if there were two cover-plates, one situated as at present, only a little thinner in section, and having a corresponding one placed on the opposite side of the joint, thus supporting the rivet at both ends, like the links of a pitch-chain, and thereby adding to its efficiency.

By referring to page 250, it will be seen that the strength of single riveting is to that of double riveting as 28 is to 35 ; hence, in the construction of the better class of boilers, the joints that have to resist the greater stress—that is, the circumferential stress—are double riveted, while those that have to resist the longitudinal or lesser strain are made with single riveting ; yet even the double rivets are relatively the weaker for their work, because the circumferential stress is doubly as great as the longitudinal stress. And when all has been done by these modifications, the joints are weaker than the original plates, to the extent of one-fifth of the strength of the plates.

With the view, therefore, of raising the strength of the joints to that of the plates, Sir W. Fairbairn introduced the system of rolling the plates with a thicker substance at the edges, where the holes have to be pierced, so as approximately to bring up the sectional area between the rivet holes to that of the general section of the plate, and thereby to obtain uniformity of strength throughout.

Another most ingenious device which has recently been proposed is, to arrange the boiler-plate joints diagonally, instead of longitudinally and transversely, according to the usual method ; by the new system the shell of the boiler is formed by winding the plates in a diagonal or spiral direction, whereby the difference of strength in the two directions is considerably modified for the better. With this arrangement the joints are at an angle of 45° with the axis of the boiler, and the strength of the joints in the weakest, that is, the circumferential direction, is thereby considerably increased.

Mr. W. R. Browne says that, 'taking the angle of the joints at 45°, and considering any square portion of the boiler surface of which the joint forms a diagonal, it will be seen that, instead of the two pairs of equal and opposite tensions, one pair double in amount of the other pair, which would act on the sides of the square in ordinary square jointing, there are, with the diagonal jointing, two equal and opposite resultant tensions acting across the diagonal joint. The resultant tension, per inch run of the joint, is found on calculation to be about four-fifths of the greater tension, and it acts not exactly at right angles to the joint, but at an angle of about 72° with it. The latter circumstance, however, does not materially affect the result, and the tension on the diagonal joint may, therefore, be taken at four-fifths of that on the longitudinal joint; consequently the effective strength of the joint and of the boiler is increased in the ratio of four to five. Thus, in a punched lap joint, if single riveted, the proportionate strength of joint as compared with that of the entire plate is increased from 55 to 69 per cent., and if double riveted from 69 to 86 per cent. The diagonal joints have, in most cases, to be replaced by transverse seams at the ends of the boiler; but as the tension on a transverse joint is only one-half of that on a longitudinal one, this does not form any objection.'

The great change which has taken place in the manufacture of plates by the rolling process, whereby they are produced of great length, breadth, and thickness, is leading rapidly to a simplification of the forms given to boilers, and to an increase of strength at the same time. The circular form of boiler is now employed for marine purposes, where rectangular boilers, with flat stayed surfaces, were formerly used, and are even yet employed in many cases. This modification, combined with elongated rivets, diagonal riveting, thickened edges, and double cover-plates, will increase the strength of the structure at its weakest points, and render the strength more nearly uniform.

The strength of a boiler is also much affected by the mode of putting the plates together, *previous* to riveting. When the holes are punched by careless workmen, it is not uncommon to find them not opposite to each other when the plates are brought together, and in order to be able to introduce the rivet, a taper-drift or punch is driven in, with violence, by means of a heavy hammer, thus damaging the line of holes to an incalculable extent, by sheer thoughtlessness. In the construction of the best class of boilers, the specification provides that the holes are to be drilled, and if any error is found to exist in the correct meeting of the holes, no drift is to be employed, any metal that has to be removed being drilled or rymered out with a cutting instrument, thus avoiding injury to the metal.

The strength of boilers depends upon the nature of the material employed for their construction: wrought iron is that which is most generally adopted, but steel is now rapidly coming into use, and in past times copper was not uncommon. Stated in even numbers, the strengths of the three kinds of material stand to each other about as follows:—

Steel	90,000 lbs.
Iron	50,000 ,,
Copper	34,000 ,,

From this comparison, it may be correctly inferred that steel will be very generally used for boilers in the future.

The strength of iron boilers is not much affected by the working temperatures, up to considerably over 400°, nor by low temperatures down to the freezing-point. But when the temperature of the plates, through the absence of water or any other cause, rises much above 500°, then a change commences. Above 750° the tenacity diminishes very rapidly, and when the plates become red hot, they have lost fully the half of their usual strength.

In those parts of steam boilers where the pressure acts upon the exterior surface of tubes, as in the internal flues of the cylindrical boiler before referred to, new conditions are established. Here the tendency is to crush, and to cause a failure by collapse of the tube. To resist external pressure the circular is obviously the strongest form, and, consequently, it is that form which is now always preferred wherever it is admissible. In such constructions, lap-joints are avoided, and welded-joints are generally substituted, although sometimes the ring is made in one or two pieces with butt-joints and cover-plates, to which the ends of the rings are riveted, either with single or double riveting.

The greatest tendency to weakness, in such flues, arises from the difference of expansion and contraction, due to the changes of temperature that are likely to arise in ordinary working, thus causing a difference of length between these flues and the exterior shell of the boiler : hence many plans have been resorted to, in order to give the different parts some freedom to act, according to their several requirements ; one of the best is that shown in Fig. 54.

This diagram represents part of the internal flue of a cylindrical boiler, such as is shown in Fig. 57 ; the tubular portions marked *a a a a* in the diagram are made by bending a plate into an overlapping circle, and then the ends are welded into a perfectly cylindrical, but short tube, of the breadth of the boiler plate. These short tubes are then joined by means of rivets, in the usual manner, to the corrugated rings or hoops, *b b*; the left-hand part of the diagram shows the exterior of the flue, while the other part shows it in section, from which it will be perceived, that the curved form of the ring gives the flue considerable liberty to expand and contract, as may be necessary, during the lengthening or shortening that takes place in the ordinary working of the boiler, but which would not be the case if the ring was made of a ⊥-iron or other solid form, or even if the flue were of one piece. This hollow form which is given to the ring likewise secures another important advantage, inasmuch as it allows the entire substance of the flue to be of nearly uniform thickness, or rather thinness, because so far as the transmission of the heat and the

Strength of Riveted Structures—Steam Boilers, &c. 257

endurance of the material are concerned, the thinner the flue, the better it is. Another advantage arises from the corrugated shape, in that it becomes a sort of circular beam to prevent the collapse of the flue.

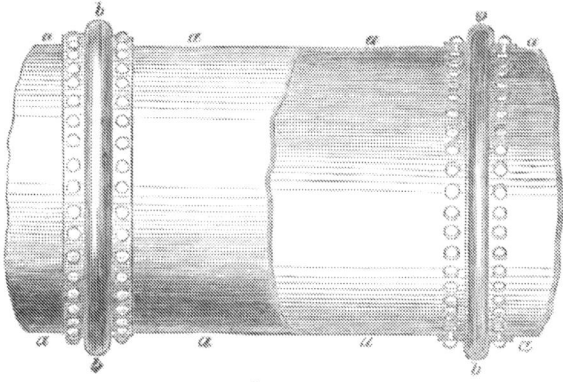

FIG. 54.

This arrangement of flue construction, therefore, not only affords liberty of motion in the direction of the length, but at the same time it adds greatly to the strength of the flues in resisting collapse, at the edge of each plate. From certain experiments, made by Sir W. Fairbairn, the marked superiority of circular to elliptical or other forms of tube, in resisting collapse, was most decided, and where strength combined with thinness are of such importance, the circular form should not be willingly departed from.

In order to make sure of the strength of steam boilers, when in ordinary use, the best practical course is to subject them to a hydraulic test, periodically, say after every 500 hours of actual work. The pressure applied should be the double of that to which the safety valve is ordinarily loaded, or one-third of the ultimate strength of the structure. There is no risk of danger arising from this test, and if only a moderate degree of intelligence is brought to bear upon the boiler, during the process, in order to watch its behaviour, all risk of an explosion from weakness is avoided, and the operation is accomplished at a cost so trifling as not to be worth regarding, when valuable life and property are at stake. Belonging to the War Department, there are about two hundred steam boilers, which are all treated as here recommended, and hitherto with the most satisfactory results.

From these general remarks, it will be seen that the strength of steam boilers depends on many conditions, but if the exact conditions are precisely known, then there will be no difficulty in estimating the strength, as due to form, dimensions, quality of material, and workmanship.

To enable the student to understand the nature of the strains in steam boilers, a few simple examples are here selected, for his guidance in the investigation of these and other similar structures.

Previous to estimating the strength of steam boilers, it is first necessary to examine and consider their shape, and the direction of the strains to which they are exposed. Let us select, as a simple illustration, the case of a plain cylindrical boiler, with hemispherical ends, as shown in Figs. 55 and 56. This boiler is exposed to rupture in two directions, as shown by the arrows, first from the ends being pushed asunder, which is resisted by the iron contained in the transverse section of the shell; second, by the bursting the cylindrical shell in the direction $b\ b$, which is resisted by the iron contained in the substance of the shell, when considered as a ring. This latter strain is similar to that in a water-pipe, and consequently the stress is as the diameter of the pipe or boiler, that is to say, a boiler of double diameter would necessarily require to have double the thickness of iron to give the same strength, whereas in the other direction, although the surface exposed is as the square of the diame-

Strength of Riveted Structures—Steam Boilers, &c. 259

ter, yet for certain reasons, which will be presently stated, it is still the stronger. Let us suppose this boiler to be 20 feet long, by 54 inches in diameter, and that it is constructed with plates of wrought iron of three-eighths of an inch in thickness. The strength in the direction $a\,a$ depends on two conditions: first, the greatest sectional area which is exposed to the pressure, namely, the diameter squared and multiplied by ·7854, which is equal to an area of 2290 square inches; second, the resistance offered by an entire ring of the iron of the shell, taken in the transverse direction, which in this case will be 54 × 3·1416, equal to a circumference of 169·64 inches, the thickness of the iron being three-eighths of an inch.

Here an important question arises, namely, What is the ultimate strength of the shell per square inch of section? The ultimate strength of the iron may be 50,000 lbs. per square inch. But the shell is weakened by the riveted joints which connect the plates together.

Suppose that the joints are double-riveted, the rivets being placed in a zigzag direction, so as to obtain the best result. Then the tenacity of the joint may be taken at 35,000 lbs. per square inch of the gross section of the plate, as has been already mentioned. For a single-riveted joint, we should have had to reduce the strength to 6,000 or 7,000 lbs. per square inch less. To allow an ample margin of safety, we will assume the strength of the double-riveted joint to be 34,000 lbs. per square inch.

The length of the ring of the shell is 169·64 inches, the thickness is ⅜ths of an inch, therefore 169·64 × ⅜ths × 34,000 will give a total resistance equal to 2,162,400 lbs.; then if we divide that strength, by the area of the section of the boiler, we shall have the pressure per square inch that would be required to burst the shell in the longitudinal direction; $\dfrac{2,162,400}{\text{area }2290} = 944$ lbs. ultimate bursting pressure per square inch.

On the Strength of Structures.

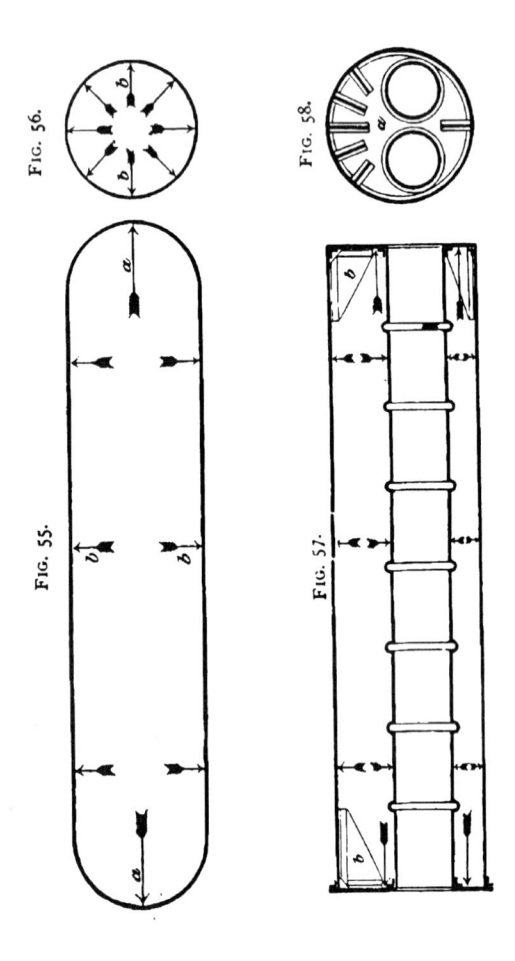

FIG. 56.

FIG. 58.

FIG. 55.

FIG. 57.

Strength of Riveted Structures—Steam Boilers, &c. 261

As regards the strength in the other direction $b\, b$, let us suppose a ring in the middle of the shell, one inch in length, this ring will be exposed to a certain pressure per inch, on a surface of 54 square inches, with an amount of iron to resist that force equal to twice $\frac{3}{8} \times 1$, or $\frac{3}{4}$ths of a square inch. Then, taking the strength of the weakest point, as before, 34,000 lbs. $\times \frac{3}{4} = 25,500$ lbs., the resistance of the ring, and this divided by the area of 54 square inches will give a total resistance of 472 lbs., which is exactly the half of the strength in the other direction. The lesson is here taught, that the shell is twice as strong in the direction $a\, a$, as it is in the direction $b\, b$, for a given uniform internal pressure.

The ultimate strength at the weakest point is thus equal to a pressure of 472 lbs., and the boiler would be perfectly safe if worked at 60 lbs. steam pressure, so that $\frac{472}{60} = 7\cdot 86$, showing that it would have a margin of safety fully $7\frac{1}{2}$ times the working pressure, which, however, is not by any means too much for a new boiler, because corrosion and other causes will soon reduce the original strength.

To select another familiar example, that of the ordinary flat-ended cylindrical boiler, with two internal flues, passing from one end of the boiler to the other end, and both riveted to the flat ends, as shown in Figs. 57 and 58. Here the conditions are widely different to what they were in the former example. In considering the extent of surface which is exposed to pressure at the two ends of this boiler, it will evidently be necessary to deduct the area of the two flues, and, still farther, in reckoning up the quantity of iron which resists pressure in that direction, it will be equally necessary to take the tensile strength of the iron, composing both the shell and the two flues, which together will place the strength of the ends in a still higher ratio to that of the shell, in a longitudinal direction, than in the case previously considered. There is one weak point, however, in these flat ends— namely, at the point marked a in Fig. 58; this arises from the flatness of the form and the thinness of the plate. If the ends were hemispherical, which is the strongest form, then it would be different. Still, for convenience, they are flat and hence weak at the point a, and therefore have to be supported, either by tie-rods passing from end to end of the boiler, or by what are termed 'gusset-plates,' bb, as shown. These are plates combined with angle irons, which are riveted both to the shell and to the upper part of the flat ends.

The course to pursue, in finding the strength of this shell, would be exactly the same as in the former example, because the shell is not assisted by the flues. Suppose, then, the boiler to be 7 feet in diameter, by 28 feet long, with two flues, each 30 inches in diameter, and for convenience let us suppose the iron to be ½ an inch thick throughout, then $\frac{34000}{84 \times 1} = 404$ lbs. as the bursting pressure per square inch of the shell. The strength of the ends will depend upon the area exposed to pressure.

$$84 \times 84 \times \cdot 7854 = 5542 = \text{whole area.}$$
$$30 \times 30 \times \cdot 7854 \times 2 = 1413 = \text{flue area to be deducted.}$$
$$\text{or } 4129 \text{ inches of surface exposed to pressure.}$$

The circumference of shell equals . . 264 inches.
Do.   do.   of the two flues . . 188 ,,
                                      452 ,,

of ½-inch iron, which is equal to 226 square inches, in strength, 34,000 lbs. per inch; then $\frac{226 \times 34000}{4129} =$ 1860 lbs. as the ultimate pressure per square inch, which would be required to rupture the boiler transversely, thus showing that this boiler is fully four-and-a-half times as strong to resist the longitudinal pressure, as it is to resist the pressure which tends to rupture the shell in the other direction.

But here it has to be remarked, that such boilers, or, indeed, boilers of any other shape, may be weakened to any extent by improper arrangements, as, for instance, by cutting out holes for the steam-dome, or manholes, without introducing material to counteract the loss of strength thus occasioned. The judicious boiler-maker will always introduce strengthening rings, or otherwise stiffen the part affected, in order to fully make up for the weakening of the general structure, due to the cutting away of the material.

There is another important practical point, which affects to a considerable extent the endurance of steam boilers, namely, the arrangement of the plates that have to be joined. This remark applies more especially to the shell and the flues of boilers of the cylindrical shape. Formerly, it was usual to unite the shell and flues to the end plates by means of an angle iron, but the better mode is to flange over the iron plate, in order to shorten the distance between the rivets and the point of exposure to pressure, and thereby to reduce the leverage of the stress which tends to open the joint. The same general remark applies to the system of overlap joints, either in the shell or the flues; there is more 'work' or stress thrown upon them than is the case when the shell is more perfectly cylindrical, with the shell-plates butted to each other, and then united by cover-plates. By this arrangement the strain is more direct, and anything that reduces the tendency to movement contributes to the permanence of the boiler.

When a boiler structure is so combined that certain parts are kept in a state of restlessness, either through differences of expansion and contraction, or by having parts of the steam engine attached to different points, all such restlessness tends to wear out the parts so exposed, not so much by the mere friction, as by a process of disintegration, whereby the incipient oxide is prematurely scaled from the surface of the plates or joints.

We may select a third example of boiler, still more complicated, namely, the tubular variety, of which the boiler of the locomotive engine may be considered as the most familiar type. As the present purpose, however, is not to explain boilers, but rather to consider their strength only, let us for convenience take this boiler in its simplest form, as consisting of a fire-box a, barrel b, and smoke-box c, as shown in Figs. 59 and 60. A slight examination of its several parts will convince the student, that its three divisions are under very different conditions as regards strength.

To select the cylindrical barrel first, because it corresponds more nearly to those boilers which have been already considered, the strength of the cylinder or barrel portion will be exactly the same as that of the shells of the two former examples, and therefore need not be further investigated. The small tubes likewise may be disregarded, because other conditions interfere, which demand for them a substance that places them far beyond the danger of collapse.

The fire-box consists of two boxes, an internal and an external, and the whole of their parts, as well as the partition-plate of the smoke-box, are all most disadvantageously placed as regards resistance. They are thin flat plates of iron or copper, which, if left unsupported, would yield with a few pounds of pressure; and when considered as a beam, with the load uniformly distributed, the strength of these plates to resist either breaking or deflection or collapse, from the great pressure of steam, may be set down as approximately *nil*, in consequence of the want of depth, due to the thinness of the plates.

To those who are not familiar with steam boilers, the suggestion may occur, that this difficulty could be overcome by making them thicker, but such a course is impracticable on account of the plates of the fire box (which is the most vulnerable part) having to transmit the heat of the furnace

Strength of Riveted Structures — Steam Boilers, &c. 265

to the water, as rapidly as possible, so that it becomes necessary to resort to other means of strengthening, in order to have the whole boiler structure of equal strength, and without requiring or employing an inordinate quantity of material to accomplish this condition.

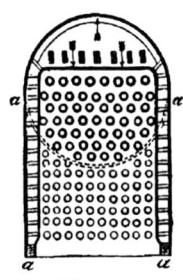

FIG. 59.

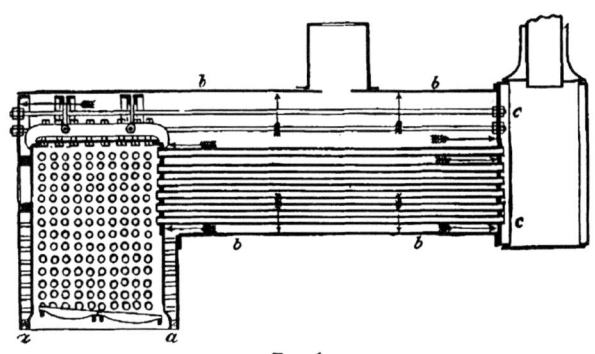

FIG. 60.

Looking at Figs. 59 and 60 it will be seen that the steam or fluid pressure is acting upon all their flat surfaces, and tending to push the inner fire-box into the furnace, and the plates of the outer box outwards. Both are equally loaded with fluid pressure, on every inch of exposed surface, and necessarily in opposite directions. This pressure, in opposite directions, is directly counteracted, and the two thin plates are united into one beam, by means of a series of copper or iron stays, these stays being repeated at every few inches of surface. The stays are inserted in such a manner as really fits them to take hold of the plates, with a grip equal to their own strength; the plates are evenly drilled through, a suitable tap is employed to screw them both, at the same time, then the stay or bolt is screwed through both, and afterwards riveted over at each extremity, thus taking a double hold of both plates. Hence, the question of strength is transferred from the plates to the stays; these stays are made of iron or copper, according to the class of boiler. Selecting the weaker metal, copper, for an example, let us suppose its ultimate strength to be 15 tons per square inch, and that we are disposed to put a stress upon it equal to 3 tons per square inch, which is as much as is suitable, and let us assume also that a stay is introduced to every 25 square inches of surface, and that each stay has an area equal to the half of one square inch, this will give us a resistance of $1\frac{1}{2}$ ton to be distributed over 25 square inches, or 134 lbs. per square inch, in addition to the modicum of assistance derived from the plates themselves. As such a boiler would not be worked at a pressure over 120 lbs. per square inch, it will be seen that the margin of safety is very considerable.

The crown of the inner fire-box is a part which is much exposed to the fire, and stands out in a solitary position away from external support. We saw, in reference to copper, p. 83, that its strength does not improve when heated, on the contrary, it becomes weaker; hence, the crown-plate, taking everything into account, is badly situated. In this strait, recourse is had to a series of thin beams on edge, or ⊥-irons variously applied, which are placed on its upper surface, and at regular distances apart, and to which the crown-plate is either stayed, or held up by ordinary rivets; the stays or rivets being of such dimensions and number as will give the required resistance. The united strength of these beams, together with that of the stays attached to the top of the outer box, must be at least equal to six times the entire pressure which comes upon the crown.

Here another difficulty occurs: the presence of such beams or ⊥-irons necessarily interferes with the passage of the heat from the fire to the water, which is not only damaging to the metal, but is disadvantageous in other respects. To obviate this objection, the beam or ⊥-iron does not rest upon the crown all over, but is separated from it, either by distance-washers, or by relieving the under surface of the beams, or by other arrangements, so that the water is in contact with a surface of thin metal, which is prevented from deflecting by stays or rivets, that derive their support from an extraneous source, and the amount of resistance may be calculated in the same manner as for the copper stays of the two fire-boxes.

From several causes, the crown of this inner fire-box is more critically situated than any other part, and is more liable to be weakened by the action of the fire; its temperature is always greater than that of the water, which is still farther aggravated by the rapid formation of steam-bubbles, which form a medium between the hot metal and the body of solid water over it, so that in determining the strength of such a vital part, all these points must be taken into account, and due allowance made for the tear and wear, during the period which it may be expected to last.

The partition of the smoke-box and the corresponding part of the inner fire-box are exposed to the same fluid pressure as the other parts, just referred to, and as they are thin they would have, if left to themselves in that position, comparatively little strength. As a rule, however, they are thicker than the other plates, not from choice but from necessity, inasmuch as they have to hold a series of small tubes, through which the fiery gases pass from the furnace to the chimney. Hence, advantage is taken of these tubes to contribute assistance to the tube plates, by offering resistance to the fluid pressure as an additional duty. Corresponding holes are formed in both plates, the tubes, nicely prepared, are securely inserted, and as additional aid, ferrules (accurately fitted) are driven into the interior of the ends, thus converting the tubes into stays.

It so happens, however, that these stays are not to be entirely relied upon, inasmuch as they are apt to work loose in the course of time. Hence, other more permanent stays of solid metal are introduced, at intervals, with nuts on the outer surface of both plates, thus rendering the structure more reliable.

From the foregoing remarks on boilers, the student will perceive that the question of strength is not so difficult as it appeared at first sight; it has only to be taken in detail, the various strains carefully worked out, and the strength of the material which is, or which has to be provided to meet the stress, may be calculated.

Resistance to Collapse.

The flues and other parts of boilers, which are exposed to the steam pressure acting from without inwards, tend to give way in another manner—by collapse. The flues are the most vulnerable part of the cylindrical boiler, unless they are properly supported, either by the corrugated rings already referred to, or by some other arrangement. The subject of collapse is so important that it is desirable that it should receive some further explanation.

This most important subject was thoroughly investigated by Sir W. Fairbairn, who carried out a long series of experiments upon wrought-iron tubes at the request of the Royal Society and the British Association, the object being to ascertain the laws which govern the resistance to collapse in such structures. The results have been freely published in Sir W. Fairbairn's 'Useful Information for Engineers.' As these experiments were extensive, the following brief remarks on the subject are only intended to describe the most important of the results which were obtained, and likewise to draw attention to some of the more important laws deduced from the experiments.

In these experiments, the tubes were firmly secured at the ends, so as to be in the same condition as flues in steam boilers, and the end fastening probably contributed in some degree to the remarkable results obtained in these experiments.

By these experiments it was shown, first, that with tubes of the same diameter and the same thickness of metal, the strength to resist collapse was nearly inversely as the length; and, second, that with tubes of the same length and thickness, but of different diameters, the strength is nearly inversely as the diameter.

In the most important series of these experiments, the tubes were all cylindrical and of the same thickness, but varied in length and diameter. The tubes were made from a single plate, bent to the cylindrical form, and the joints riveted and brazed to render them water-tight; the ends of the tubes were riveted to rigid cast-iron discs, and the air from the interior of the tube was allowed to pass out, when the tube collapsed, through a pipe screwed into one of the discs. The tubes were placed in a strong cast-iron cistern filled with water, and the pressure was applied by a hydraulic pump, and its amount was registered by two gauges of the best description. It is to be noted that the term collapse, as here used, does not imply the crushing of the material, but simply that the circular or elliptical arched shape of the transverse sections changes its form, at the weakest part of the tube, by the thin plates either bending or wrinkling. The results of this set of experiments are shown in the following Table:—

On the Strength of Structures.

Results of experiments made to ascertain the resistance of cylindrical tubes of wrought iron to collapse when the ends are firmly fixed, and the tubes are subjected to an external pressure.

TABLE XLIV.

1	2	3	4	5	6	7
	Size of the tubes.			Mean pressure in lbs. per square inch at which the tube collapsed.	Comparative numbers showing the collapsing pressure reduced to unity of length and diameter.	Mean of comparative numbers.
Number of the experimental tube.	Diameter in inches.	Length in feet.	Thickness in decimals of an inch.			
1·2·6 3 4 5 27 29	4	$1\frac{7}{12}$ $3\frac{1}{4}$ $3\frac{3}{8}$ 5 5 $2\frac{1}{2}$	·043	149 65 65 43 47 93	930 866 823 860 940 930	891·5
10·11 9	6	$2\frac{1}{2}$ $4\frac{1}{12}$	·043	58·5 32	877 944	910·5
13 14 15	8	$2\frac{1}{2}$ $3\frac{1}{4}$ $3\frac{1}{3}$	·043	39 32 31	780 832 826	812·6
16 17	10	$4\frac{1}{8}$ $2\frac{1}{2}$	·043	19 33	791 825	808
18 19 20	12·2 12	$4\frac{3}{4}$ 5 $2\frac{1}{2}$	·043	11 12·5 22	654 750 660	688
22 23 24	18·75 9	$5\frac{1}{12}$ $3\frac{1}{12}$ $3\frac{1}{12}$	·25 ·14	420 262 378	40031 7270 10489	

It has to be noted that the experimental tube No. 6, although it was really 5 feet long, yet had two rigid cast-iron rings soldered upon it, dividing it into three equal parts, 19 inches in length between the rings, and hence this tube is placed in the table with the tubes whose real length was 19 inches. The result shows that, although the tube was really the same length as Nos. 5 and 27, it yet possessed three times the strength to resist collapse, a result which was due solely to the tube being kept in the true circular form by the cast-iron rings, and thus made virtually into three short tubes instead of one long tube.

There are two great lessons taught by this set of experiments. 1st. The resistance of tubes of uniform thickness varies inversely as the length, or nearly so. This law is clearly shown by the experiments, and by comparing the 4-inch tubes Nos. 5 and 27 with No. 29. It will be seen that the former, which are twice the length of the latter, collapsed with about half the pressure, and Nos. 1 and 2, which were rather less than one-third the length of Nos. 5 and 27, bore rather more than three times the pressure.

The 6-inch tubes also follow the same law; Nos. 10 and 11 collapsed at 58·5 lbs. pressure, and No. 9 should have borne a proportionate pressure, as due to its length, which may be calculated thus ;—59 inches : 30 inches :: 58·5 lbs. : 29·74 lbs. But the actual collapsing pressure of the latter tube was by experiment 32 lbs., so that the difference between the calculated and actual pressure was only 2¼ lbs. per square inch, which is very slight.

The resistance of the 8-, 10-, and 12-inch tubes also bears the same relation to the length, and hence it may be concluded that the strength of such tubes, to resist an external pressure, varies inversely as the length, at least between the limits of 1½ and 10 feet.

The furnace or flue tubes of steam boilers, at the time of these experiments, were made without any support between the extremities, but an alteration speedily followed, so soon as the knowledge gained by these experiments was made known to the world. At the present time, the tubes of all well-constructed steam boilers are strengthened by rings of a T section, or of a corrugated section, these rings being riveted to the tube at intervals of about 3½ feet, and this has added greatly to the security of boilers.

The second lesson taught by these experiments is, that the resistance of such tubes to collapse, when the length and thickness are the same, varies inversely as the diameter. As an example, take No. 29, a 4-inch tube 2½ feet long, with a collapsing pressure of 93 lbs. per square inch, as a standard. Then, by calculation, the 6-, 8-, 10-, and 12-inch tubes of the same length and thickness should collapse with pressures of 62, 46·5, 37·2, and 31 lbs. per square inch respectively. By experiment, the pressures were found to be 58·5, 39, 33, and 22 lbs. per square inch.

The column No. 6 is composed of numbers obtained, by multiplying the collapsing pressure of the tubes by the length and diameter of the tube, to show how nearly the resistance of tubes of the same thickness is inversely proportional to their diameter and length; and, had the tubes followed this law exactly, the numbers in that column would have been equal. The greatest discrepancy occurs with the 12-inch tubes.

The differences in these numbers, and likewise between the calculated and experimental collapsing pressures, may be accounted for, by the varying strength of the material and differences of workmanship, and, in the case of the 12-inch tube, by the difficulty of making so large a tube of such thin material, so that it may be perfectly cylindrical in all parts.

The experiments Nos. 22, 23, and 24 in this Table were made with three tubes having a greater thickness in proportion to their diameters. Two of these tubes were precisely similar in all respects, but that the one was made with a 'lap' joint (Fig. 61), and the other with a 'butt' joint and external cover plate (Fig. 62).

Strength of Riveted Structures — Steam Boilers, &c. 273

These experiments were made to ascertain, first, to what extent the strength of the tube was reduced by the slight departure from the true circular form, which is unavoidable when a 'lap' joint is used; and secondly, to ascertain, if possible, the different powers of resistance of thick tubes of different diameters. The result of the experiments shows that a loss of rather more than one-third of the strength was caused, by the slight deviation of a quarter of an inch from the true circular form in the 'lap'-jointed tube; this should be carefully noted.

In well-constructed boilers, any deviation from the true form is avoided by making the tubes in short lengths, and

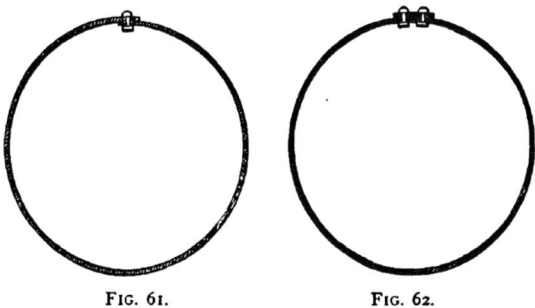

FIG. 61.   FIG. 62.

'lap-welding' the joint, care being taken that the thickness of the tube is maintained, but not exceeded, at the weld. The cylindrical form is further secured by the strengthening rings, which are now rolled out of the solid 'bloom,' and are therefore weldless and perfectly cylindrical, so that any departure from the proper cylindrical form is inexcusable.

The resistance of the above three tubes was found to be much greater than that of those having a less thickness in proportion to their diameter, and Sir W. Fairbairn concludes that the strength of tubes of the same length and diameter,

T

but of different thicknesses, varies as the 2·19th power of the thickness. For all practical purposes and to facilitate calculation, the square of the thickness may be used instead of the more correct 2·19th power, so that with two tubes precisely similar in every respect, but that the one is twice the thickness of the other, the strength of the former to resist collapse will be very nearly four times that of the latter.

In all the foregoing experiments, the ends of the tube were securely fixed, and further experiments were made to ascertain whether similar tubes follow the same law, when the ends are left perfectly free. Two tubes were used for these experiments, they were 8 inches diameter and 30 and 60 inches long respectively, the result is shown in the following Table :—

Result of experiments made to ascertain the resistance of cylindrical tubes of wrought iron to collapse, when the ends are perfectly free to approach each other, and the tubes are subjected to an external pressure.

TABLE XLV.

Number of experimental tube.	Size of tubes.			Pressure in lbs. per square inch at which the tubes collapsed.
	Diameter in inches.	Length in feet.	Thickness in decimals of an inch.	
25 26	8	5 2½	·043	22 36

It appears from this Table that the strength is not inversely as the length, as was the case when the ends of the tubes were fixed, or the shorter tube should not have collapsed until the pressure reached 44 instead of 36 lbs. per square inch.

Comparing these experiments with No. 13 in Table XLIV., it will be seen that a similar tube to No. 26, but with the ends fixed, collapsed with 39 lbs. pressure, so that it appears that the freedom of the ends to approach each other does not decrease the strength to any considerable extent, provided the circular form is still retained at the ends of the tube.

The next experiments were made with tubes of an elliptical form of section ; the result is shown in the following Table :—

Result of experiments made to ascertain the resistance of elliptical tubes of wrought iron to collapse, compared with that of cylindrical tubes, when subjected to an external pressure.

TABLE XLVI.

Number of the experimental tube.	Size of tube.			Pressure in lbs. per square inch at which the tube collapsed.
	Diameter in inches.	Length in feet.	Thickness in decimals of an inch.	
34	$14 \times 10\frac{1}{4}$	5	·043	6·5
35	$20\frac{3}{4} \times 15\frac{1}{2}$	$5\frac{1}{12}$	·25	127·5

The tube No. 34 was of the same thickness, and had the same quantity of material in its section, as the cylindrical tube No. 19 in Table XLIV. ; the latter collapsed with a pressure of 12·5 lbs. per square inch, and the former with a pressure of 6·5 lbs. per square inch, thus showing that a loss of strength equivalent to a pressure of 6 lbs. per square inch, or 48 per cent., was caused by flattening the tube, so that its least diameter was reduced $1\frac{3}{4}$ inch, although the quantity of material in both tubes was exactly the same.

The tube No. 35 was of the same thickness as No. 22 in Table XLIV., and had only $\frac{3}{8}$ths of a square inch less material in its section, but it collapsed with a pressure of 127·5 lbs. per square inch or 272·5 lbs. less than the cylindrical tube, showing a loss of strength of 64·8, or nearly 65 per cent., by flattening the tube until its least diameter was $\frac{1}{4}$th less than that of the true cylinder.

Tubes exposed to Internal Pressure.

A few experiments were carried out at the same time to ascertain whether the length of tubes similar to some of those previously enumerated, affected their resistance to an internal pressure. The results were, however, far from satisfactory, only two of the five tubes experimented upon being sound at the joint. These two were 6 inches diameter and 12 inches and 48 inches long, respectively; the former burst with a pressure of 475 lbs., and the latter with 375 lbs. per square inch, apparently showing that the resistance varied in some degree with the length; but when practically considered, the result would appear to show that the length of the tube does not affect its strength under such circumstances, and both tubes would have burst with about the same pressure, had not the shorter one, owing to its extreme shortness and the consequent proximity of the point of fracture to the end supports, derived a great amount of its resistance from the fixing of the ends.

Some further experiments were then carried out with leaden tubes of 3 inches diameter, and $14\frac{1}{2}$ inches and 31 inches in length, respectively. These tubes were of the same thickness of metal and burst with a pressure of 374 and 364 lbs. per square inch respectively, from which it may be inferred that the shortness of a tube does not contribute to its resistance to an internal pressure, unless it is extremely short, say less than two or three times its diameter in length, when its apparent strength would be partly due to the support given by the ends of the tube.

In the Perkins system of steam boilers, the boiler is composed of small tubes, which sustain with safety a pressure of 5,000 lbs. per square inch. To obtain greater economy, steam pressure increases from year to year; it is therefore probable that the steam boilers of the future will be gradually modified, by the more general employment of small tubes.

CHAPTER XVII.

ON STRUCTURES SUBJECT TO INTERNAL PRESSURE.

THE present chapter will treat of the strength of cast-iron pipes and water tanks, and also of the relative advantages of hooped and solid cylindrical structures exposed to internal pressure, as, for instance, in the case of guns and hydraulic press cylinders.

Cast-iron Pipes.

Cast-iron pipes are now so extensively used for various purposes, that it is of importance for the student to examine their power of resistance.

In the majority of purposes for which cast-iron pipes are employed, as for the conveyance of gas or of water for the supply òf towns, or for steam pipes or steam cylinders, or even for high-pressure hydraulic pipes, the question of strength is not so important as might be inferred from the trouble experienced in keeping them sound and water-tight. They are usually cast of such a thickness that their strength is apparently in excess of that required to resist the pressure acting on them. This superabundance of strength is given, because other practical considerations step in, which, as a rule, render it absolutely necessary to make such articles considerably stronger than is actually required for the work which they have to do. The practical considerations here referred to are the limits to thinness and soundness attainable by the founder, and the comparatively fragile nature of cast-iron pipes, in bearing the rough handling to which they are subject in transit, and more especially the straining due to the subsidence of the earth from under them when laid underground, or to its compression during the passage of heavy waggons over them.

Reckoning the tenacity of cast iron of the quality which is frequently used for common pipes at 16,000 lbs. per square inch, it will be found by calculation, that very thin pipes of such iron would resist a great water pressure. Let us take a ring cut from a pipe, with a bore of 10 inches in diameter, and suppose the said ring to be 1 inch in length and ½ an inch in thickness, as in Fig. 63 ; this will give a

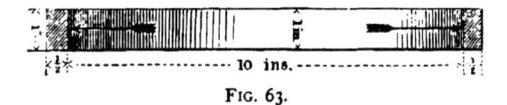

FIG. 63.

substance of iron equal to one square inch of section to resist the water pressure. But one square inch will have an ultimate strength of 16,000 lbs., and may in such a case be safely strained to 4,000 lbs. To produce that stress, the pressure in the pipe must be 4,000 ÷ 10, or 400 lbs., per square inch, or above 26 atmospheres. Hence, it will appear that, but for the reasons already stated, gas and ordinary water pipes might be much thinner than they are usually made.

In the Belgian Annexe of the Paris Exhibition of 1867, a cast-iron pipe was shown, 20 feet long and 28 inches in diameter, and varying from ¼th to ¾ths of an inch in thickness. This pipe was proved with a pressure equal to 5 atmospheres. In this case the stress put upon the iron at the thinnest part of a ring, one inch in length, would be as follows :—The pressure of 5 atmospheres would be 75 lbs. per square inch, the number of inches being 28. The total substance of the two sides of the iron ring would be equal to ½ an inch. Then 75 × 28 will give 2,100 lbs. as the total pressure of the fluid which had to be resisted by the ½ inch of iron, or equal to a stress of 4,200 lbs. per square inch, thus only straining the iron up to the quarter of its ultimate tenacity; such a pipe, however, would be easily broken, unless great care was exercised in handling it.

The water pressure pipes employed in connection with hydraulic crane works, are seldom used under a pressure of less than 700 lbs. per square inch, and the pressure varies from 46 to 68 atmospheres in different cases. The pipes conveying the water, when 3 inches in diameter, are made with ⅝ths of an inch of thickness, and are proved with 2,500 lbs. internal pressure per square inch. Taking a ring of this pipe 1 inch in length, the water pressure tending to burst it will be 2,500 × 3 = 7,500 lbs., which has to be resisted by ⅝ × 2 = 1¼ square inch of iron, which, for such a purpose, would be of rather a better quality than that for common pipes, probably having a tenacity of 18,000 lbs. per square inch. This would give an ultimate resistance of 22,500 lbs., or three times the stress to which the ring is exposed, under proof, and the proof stress is fully three times greater than the stress during ordinary working. It might be inferred that such pipes were unnecessarily strong, but such is not the case; owing to the numerous contingencies to which they are exposed, and to the effect of continued rusting while buried in the earth, they require an excess of strength, and experience fully confirms the wisdom of allowing a large margin at the outset.

Cast-iron Tanks.

In the construction of round, square, or rectangular tanks, built up of cast-iron plates, which are united with wrought-iron bolts by means of flanges or ribs, which are cast upon the edge of the plate at right angles to it, different conditions exist to those met with in pipes which are of comparatively small diameter. Let us select a round water-tank as an example, which may be compared to an immense tube placed upon its end. The enlargement of the pipe into a water tank of say 100 feet in diameter by 25 feet in depth, brings the question of strength into greater prominence. It is a very common error to suppose that the strength of the tank is not affected by the diameter, but, as we shall see, the diameter has a most important influence on the resisting powers of a tank.

In the first place, a pressure or weight of 25 cubic feet of water, equal to 25,000 ounces, rests upon every square foot of surface of the bottom plates, but as these are here supposed to be lying upon a solid foundation, the strain on them may be disregarded.

The pressure per square foot that comes upon the sides of the tank is only equal to the half of that which rests on the bottom, because at the surface the pressure is nil, while at the bottom it is, as before stated, equal to 25,000 ounces ; consequently the fluid pressure due to $12\frac{1}{2}$ feet, or 12,500 ounces, is the average pressure that the side or circumference of the tank has to sustain, per square foot of surface.

The above remarks apply to every kind of tank, but it would be a great waste of iron to make the upper tier of plates, in a tank, sufficiently thick to withstand the pressure of water at $12\frac{1}{2}$ feet in depth, and it would be still worse to make the lower tier of plates only equal to the pressure of $12\frac{1}{2}$ feet, seeing that they have to withstand 25 feet. This renders it obvious that, in order to obtain the requisite strength with the minimum of iron, the bottom tier must be equal to the strain that comes upon them, and the thickness upwards must be gradually diminished to the top plates, which need only be of such a substance as will meet the various contingencies referred to, in connection with the casting and conveyance of pipes.

If we select one foot of the lowest ring of iron, and assume the average pressure upon it to be 25 feet of water, and then consider it as similar to an inch ring of a pipe, we shall then see that the question of strength is important. The iron used for such a purpose would probably have a tenacity of 16,000 lbs. per square inch ; then we may inquire how much of such iron will be necessary to withstand the pressure, leaving out of consideration, for the present, any assistance which it derives from being fixed to the edge of the bottom of the tank.

The tank is 100 feet in diameter, which, multiplied by 25,000 ounces, the water pressure on every foot, is equal to a total pressure tending to burst the lower ring, equal to 2,500,000 ounces. Suppose we resolve not to strain the iron above ⅓rd of its ultimate tenacity ; ⅓rd of 16,000 lbs., when reduced to ounces for convenience, is equal to 85,333—say 85,000 ounces. Then, dividing the total water pressure on the ring, tending to tear it open, by 85,000, will give 29·4 as the number of square inches of iron required (say 30 inches); that is to say, the two sides of the ring should have between them 30 inches, and, each being 12 inches in length, the thickness would consequently have to be 1¼ inch. If the tank is considered merely as a pipe, then 1¼ inch of thickness would be required, but if we take into account the assistance derived from its connection to the bottom plates, as well as the support due to the projecting flanges, which act as ribs, then it will be seen that the 1¼ inch may be reduced to 1⅛ inch, with perfect safety. In the same way will have to be calculated the number and size of the wrought-iron bolts that are required to hold the plates together, and which ought to be at least equal in strength to the 30 inches of cast iron.

In order still farther to reduce the substance of the plates and the cost of the tank, it is now customary not to rely entirely upon the cast iron, but to reduce the substance of the lowest plate still farther—say to 1 inch in thickness— and to make up the difference by wrought-iron bands or hoops, which are put on under tension, and which, by their greater tenacity and reliability, afford the necessary security. By such combinations, the practical engineer is in some measure enabled to attain the maximum of strength and minimum of cost.

In the foregoing remarks on the strength of pipes which are comparatively thin, it has been assumed that, when they are exposed to internal pressure, the metal composing the pipe is all performing duty in its resistance, in the same manner as a bar of iron when pulled asunder by tensile force. This is not, strictly speaking, the case, although in large structures, such as the water-tank, it is very nearly accurate to assume the metal to be uniformly strained; and, even in small water pipes, no error of practical importance is introduced by that assumption. In thick pipes, or in cylinders for hydraulic presses, or in gun structures, it is, however, widely different; then it becomes imperative to treat the question in another manner.

On the Strength of Hooped as compared with Solid Structures.

It was pointed out by Professor Barlow, many years ago, that the strength of a pipe, hoop, or cylinder to resist internal pressure is not in proportion to the mass or thickness of material of which it is composed, and that by adding to the thickness of a tube or a gun, or to the substance of the cylinder of a hydraulic press, or by increasing the thickness of an iron pipe, we do not thereby increase the strength, in proportion to the quantity of metal which is thus added. When a hollow cylindrical vessel is exposed to internal pressure (amounting in hydraulic presses to 3 tons per square inch of surface, and in guns to a still greater intensity) the pressure does not affect the whole mass of metal opposed to it in the same degree. The metal composing the inner surface of the bore is first affected, and as that part is extended, so the stress gradually reaches to the next lamina, but in a decreasing ratio, and this transmits it to the next, and so on, until at length the resistance of all are in some degree brought into active exercise.

The important point to observe is, that the work which is

Structures subject to Internal Pressure. 283

performed by the successive concentric laminæ, in resisting the internal pressure, is in exact proportion to the amount of stretching to which they are severally subjected. It is found by experiment that, when a ring is stretched by mechanical means, the outside does not expand so much as the interior; this will appear to be self-evident, because if it did so, then the volume of the material composing the ring would actually be increased thereby, which is impossible; but besides and independently of any experiment, as the outer circumference of the ring is necessarily longer than the inner surface, even if they did stretch an equal amount (if that were possible by a narrowing of the ring), the metal composing the two surfaces would not even then be strained equally, on account of the stretch on the outside being distributed over the circumference of a larger ring than that on the inside. And upon the above conclusion some most important practical deductions rest.

In attempting to design a gun or cylinder, which shall be theoretically perfect, so that the whole of the molecules which compose the cylindrical part of the structure shall stretch alike, then the metal should be so disposed that, at the moment of pressure or explosion, every part thereof shall be equally ready to take its full share of the duty at once, and in proportion to its ability, and should be constructed with the tension previously put on the alert, so as to be always in readiness, and without the outer portion having to wait for the stretching of the interior to give it employment.

Up to the present time, no practical system of constructing either guns or hydraulic cylinders comes nearly up to these theoretical conditions; the nearest approach is probably attained by building up the cylinder with fine wire, which is wound round an interior barrel, the wire being put on under definite tension; but such an arrangement, although fulfilling one set of conditions, would evidently be weak in the longitudinal direction of the gun, unless all the wires were soldered together into a homogeneous mass, and even then the result would be doubtful.

Another approximation to the theoretical conditions is secured, in the American system of casting iron guns, by cooling the mass of hot metal from the interior of the bore, instead of by the usual method of allowing the mass to remain in the foundry mould, until the heat has passed away by conduction through the exterior. In the latter case, it may be inferred that the outside of the mass of metal is necessarily colder than the centre of the solid block, and consequently will have contracted more on that account, and so have become stretched upon the warmer and more expanded interior, until at length the whole of the heat has passed away, and the inside consequently has become equally cooled, which will then likewise have contracted and taken up its normal dimensions, and will thus find itself at a disadvantage on account of the previous stretching and final setting of the outer portion. It will then necessarily become less dense, and, hence, to some extent, it will lose the full grip and support of the exterior metal composing the mass.

By the American plan, on the other hand, the gun is cast hollow on a mandril, with what is termed a water-core—namely, a tube in which a constantly circulating stream of water is kept up, to carry off the heat rapidly from the centre, thus entirely reversing the conditions, and so causing each lamina in succession to grip hard upon that which it encloses.

In carrying out this cooling operation, shortly after casting, the stream of water is first directed through the core-barrel, entering by a pipe down the centre and then rising through the annular space between the pipe and the core-barrel, and escaping by passing away over the top of the dead head. In this process of cooling a heavy gun by a stream of water, the procedure may have to be continued for a couple of days, and when the gun has partly cooled, the core-barrel is extracted, and the water is made to flow into the bore by a pipe, in a similar manner as before, but the water now escapes in contact with the actual bore of the gun; meanwhile a fire is kept up all round the exterior mould of the gun, in order to protract the high temperature of the outside, and prevent the heat from escaping in that direction.

Independently of the foregoing consideration, the question of strength in a solid cast-iron cylinder or gun is likewise affected by the exterior form, the presence of the gun trunnions, or, indeed, of any other massive projections or irregularities of outline, and still more by the shape given to the breech. All these points determine strength or weakness, inasmuch as these forms or shapes create new conditions, which greatly affect the direction in which the waves of heat pass out, from the interior of the mass, to the nearest point of exit into space. It is found that all such irregularity introduces elements of discordance, into that which would otherwise be the harmonious and natural order of crystallisation, any departure from which is always accompanied by corresponding weakness.

A good illustration of this kind of weakness was afforded by the accident which occurred in raising the Britannia Tubular Bridge. The hydraulic cylinders were originally made with a flat bottom, like that of a drinking-glass (as shown in Fig. 64); the cylindrical part of the casting had the crystals radial from the inside, but in the bottom part the crystals were perpendicular to the flat end, and at the points where the two different arrangements of crystals come together, at an angle—namely, at a line drawn from the inner to the outer corners—there were the lines of weakness. Hence it was that, although considerably thicker, the cylinder failed at those points.

The second or substitute cylinder was made with a hemispherical end (as in Fig. 65), and in it the radiating crystals were all arranged in lines more nearly parallel, although of course not truly so. As thus made, it was found to be amply sufficient in strength, even with the same amount of metal in the mass. The foregoing accident was the means of drawing public attention to the subject. Many engineers found it to agree with their former experience, and many

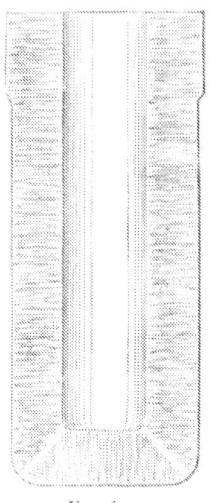

FIG. 64. 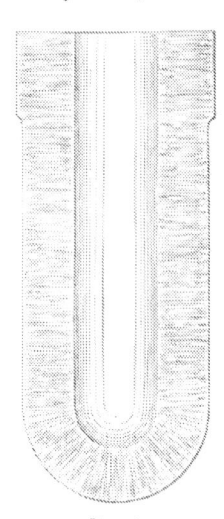   FIG. 65.

had previously been modifying the forms of structures without knowing the natural law.

The subject is well illustrated by a singular phenomenon that is observed in chilled shells, in which, from the effect of the sudden deprivation of heat, the lines of crystallisation are very clearly marked, and, when broken, the order of crystallisation is exhibited most convincingly. In such shells there occurs an apparent inconsistency with this reading of the law, namely, at the point of the shell, where in every case the crystals take a curved direction rather than straight out into space (as shown at Fig. 66). This is due to the following cause:—The shell is cast on end, with the point downwards. When the liquid metal is poured into the mould, it is, as a matter of course, in contact all over; and during the period that it remains a liquid, it naturally follows the gradually expanding mould or vessel in which it is contained; but as it begins, from the effect of the chill, to form an exterior crust, the time arrives when the body of the casting is not in actual contact with the mould, and consequently the rate of conduction of heat is thereby lessened, but not so with the point of the shell, which is kept at the bottom of the mould by the effect due to the gravity of the mass. Hence, the point of contact becomes a new line of direction, competing with the sides of the mould, and the effect of the two determines the curved form of the crystals, as shown in every instance.

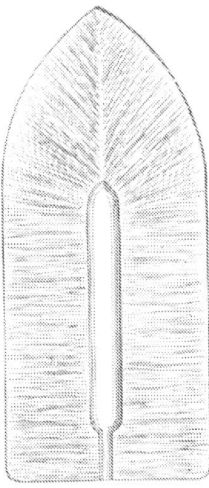

FIG. 66.

In past times, it was assumed that cast iron must necessarily be homogeneous, instead of which it is otherwise, any sudden divergence of the escaping lines of heat causing a change in the direction of the crystalline formation, and rendering that part weaker than the general mass; and as with a chain so in a gun or cylinder, the strength is only equal to that of the weakest point. The knowledge and right application of this law will in time affect the form of many structures, and lead to the use of cast iron for purposes where wrought iron is now employed.

Already most of our engineers are constructing their hydraulic cylinders on the plan shown in Fig. 65, and the Americans are constructing their cast-iron guns in form not unlike a soda-water bottle, which is nearly in strict accordance with this law of crystallisation, the exterior surface being arranged so as to invite the heat outwards in a nearly uniform current in all directions.

Built-up Guns.

We have already indicated that strength is derived from putting the interior mass of a cast-iron gun or cylinder under compression, by the initial tension of the metal nearer to the outside, due to cooling from the interior, and thus approximating to the condition of the gun made with fine wire. But it has further to be observed that the same conditions would be obtained if the structure were composed of a great number of thin hoops, put on one over the other, but under the same tension as the wire, if this were practicable, which it is not commercially, on account of the expense in producing such accuracy as would ensure the specified tension. But although it may not be convenient to construct such articles with a great number of thin hoops, still, by a slight departure from the theoretical conditions, namely, by using a smaller number of thick hoops, an approach may be made to great strength, and by thoroughly practicable means. In this way we can fabricate large guns or cylinders, by taking the several parts in detail, and combining them into a comparatively perfect whole, at a moderate expense, and with such an approximation to the theoretical conditions as affords great satisfaction; the great condition aimed at being to equalise the stress throughout the mass, when under tension from internal pressure, and this is the principle of the system now pursued both at Woolwich and Elswick in the construction of what are termed 'built-up guns.'

The wrought-iron hoops are made of any diameter or length, by first making a bar of the required section, and of sufficient length to contain the necessary quantity of iron; this long bar is then heated in a furnace, and by mechanical means it is wound round a mandril into the condition of a coiled bar. This coil is removed from the mandril and then put into a furnace, made welding hot, and then put under a steam hammer, and the loose coil is then welded into a close cylinder or hoop of wrought iron. Or, the above process may be varied, in the formation of exceptionally thick hoops, by the winding of one coil over another, before the welding operation is performed. When the hoop is forged, it is put into the lathe or other machine, for boring and turning it to the required dimensions.

On the first consideration of this subject, the mind is rather unwilling to believe that in such articles as are here referred to, when composed of fine wire or a number of thin rings or hoops, or even of a smaller number of thick hoops, as used in the construction of modern built-up guns or cylinders, that such a mode of construction can have the same solidity and strength to resist internal pressure, as a similarly formed homogeneous mass in one forging or casting; still more so, when that mass is made of the best material, and in practice shows great endurance, as is the case with the fine cast-steel guns made by M. Krupp of Essen. Besides, it might further appear that in the built-up structure, when consisting of a hooped fabric, in which the mass composing it is discordant, being neither homogeneous nor working in harmony—that each hoop is under different conditions, that the barrel and the inner hoops are existing under compression, while the exterior hoops are under great tension. Nevertheless, the positive results show that no sensible practical disadvantage is found to arise from the want of homogeneity, while the homogeneous or solid guns are not more reliable, if indeed they are equally reliable. By the building-up system, we are enabled to have the finest steel for the interior barrel, and a cheaper material, wrought iron, for the remainder; and although wrought iron is a cheaper and, in one sense, an inferior material, yet for this special purpose, from its great toughness, it may, as a general rule, be considered fully equal, if not superior, to steel of greater ultimate tenacity.

From these remarks, it will be seen that the subject admits of a difference of opinion, and such a difference does exist amongst those well qualified to judge, and who have given the subject great attention; but still the fact remains that the strain which comes upon the interior of a gun first acts on the inside of the bore, and as that part of the metal becomes stretched, so the stress gradually reaches the outside in a diminishing degree, and hence the outer metal cannot contribute its full share of duty, unless it has an initial tension.

This principle of initial tension is employed, in the modern manufacturing system of building up wrought-iron and steel guns and hydraulic cylinders, by the various modes of either shrinking or pressing one hoop, under tension, over another hoop.

The shrinking system, which was introduced by Sir William Armstrong, for the construction of artillery, has been extensively applied, and at the present time it appears likely to supersede all other systems, from the circumstance that, by this simple arrangement, the best practical results as to quality and cost have been obtained.

After taking all things into consideration, greater practical advantages have probably been attained by this system than by any other, and as the same principle of construction is equally applicable to hydraulic cylinders, the student should endeavour to understand its general bearing.

To take the gun as an example, the principal part is the bore, formed within the inner barrel, which constitutes the foundation or core upon which the exterior hoop structure is to be built up. The chief object of the hoops is to render all the support of which they are severally capable,

Structures subject to Internal Pressure. 291

by the principle of having the portion of work that they can do, partly put upon them in the original making of the gun. The inner barrel is, then, the most important part; it is made of cast steel of a mild quality, and carefully tempered in oil, so as to bring out its strength and elasticity. It is then covered with a thick hoop, or a series of thick hoops in succession, the one over the other, and each under regulated tension, as determined by calculation of the relative dimensions of the parts that have to grip and the part which has to be gripped, and they are so combined into a whole that each hoop shall take its full duty at the critical moment, and that the total tension put upon the several hoops shall be at least equal to or exceed the effect of the explosion.

The possibility of attaining such conditions is due to the knowledge of the fact that iron stretches about the $\frac{1}{10000}$th part of an inch, per inch of length, by a ton of stress. It therefore becomes a simple matter of calculation, to determine the difference of dimensions between the outside of the inner surface and the interior of the outer that will give the required tension, and consequently the stretch and specified grip in tons of positive support.

When the two surfaces are made to the required diameters, then the outer hoop is heated to redness, which causes it to expand about $\frac{1}{100}$th part of its linear dimensions; it is then carefully slipped over the inner part, upon which it gradually cools and contracts, with the force, in tons per square inch, upon the metal, intended to be applied.

As the greatest stress or pressure from the explosion will be brought to bear, first, upon the interior of the inner barrel, and from that will be passed on to the hoops, the object is to concentrate the grip of all the hoops upon the inner barrel. The effect of the united grip of a series of hoops, due to shrinking one hoop over another, is, that the grip of an outer hoop serves partly to undo a portion of the grip of the hoop or hoops which are under it, thus putting the entire structure into a state of lively activity; and the several hoops are always standing at attention, to perform the required duty; but in determining the precise dimensions which will assign the proper tension and duty to each hoop, all the alteration or disturbance of dimension, and consequently of tension, thus caused by the successive grip of hoop over hoop must be taken into account, in establishing the difference of dimension between the inner surface of the gripping hoop and the outer surface of the under hoops, as also the relative grip to be put upon the hoop next to the barrel, and the intermediate and outer hoop.

In the attempt, so far as it may be practicable, to place the whole structure in such a condition of lively tension as that, at the moment of explosion, each hoop in succession will be already under the assigned load, and all of them take the same duty in tons per square inch, it is necessary that the amount of tension put upon the several hoops should not be equal, but should be inversely as the stress that would reach them severally, if the structure were a solid mass of metal.

Until recently it was generally considered, on theoretical grounds, that the force or resistance exerted by the different parts of a solid cylinder, was inversely as the square of the distance of the parts from the centre, but recent experiments, made by stretching rings within rings, would seem to prove that the stress is nearer to the inverse ratio, or at least nearer to that than to the former. The cause for any such uncertainty is owing to the physical nature of materials, and arises from a complication of various reasons due to the properties of the metal—its elasticity, ductility, and compressibility, which all step in to interfere, complicate, and introduce uncertainty, and all come into active play and thereby affect the apparent result. But all experiments point in the one direction—namely, that the strain comes upon the interior in all its force, and gradually decreases inversely as the distance from the centre, or nearly so. Therefore when a gun is completed, the tension of the structure should be greatest at the outside, and gradually decreasing to the inner barrel, which should be under compression, so that at the moment of greatest stress no part should be strained beyond the apparent limit of elasticity.

To those who are not familiar with the practical working out of this system of building up concentric structures, it might appear that some difficulty would be experienced, in obtaining such accuracy as would ensure the required degree of tension, with uniformity; but such is not the case, from the circumstance that the system of manufacturing has been so organised as to reduce the results of measurement to a certainty. The hoop is first bored to the intended dimension; it is then carefully measured by instruments, that by the aid of the vernier or micrometer read off the dimensions to the $\frac{1}{10000}$th part of an inch. As the interior surface may not be perfectly parallel, the correct size is noted at different points and written down on paper. To these dimensions are added the amount of difference required (namely, $\frac{1}{10000}$th of an inch, per inch, for each ton of tension or grip). The total dimension gives the size of the part on which it has to be shrunk. These new dimensions are recorded, and with the corresponding measuring instruments duly marked in ink, and the same dimension written upon a sketch; the paper and guages pass on to a turning-lathe, where the respective parts of the structure are reduced to the required diameter, an operation necessarily requiring skill and care on the part of the workman, but no difficulty is experienced.

Should the operation be performed erroneously, by making the grip too small, then it is only necessary to reserve the hoop for the next gun, and select a smaller hoop for the one in hand, these differences being only a few thousandths of an inch. If, on the other hand, too much metal has been left on the exterior diameter, thus creating a greater tension upon the metal composing the hoop than was intended, the error will be detected by the system of check, when it reaches the official viewer at the next stage, and is then either passed or returned by him for correction by the turner.

From some observations which were made on the above point at the early stage of the manufacture, before much experience had been gained in regard to the shrinking of exterior hoops, it seemed more than probable, from certain results, that the strength of the hoop was not much, if at all, imperilled by too much tension being put upon it; that the iron hoop, from being hot, probably allowed itself to be stretched or permanently wire-drawn to a certain extent without injury, and thereby was enabled to accommodate itself to a wrong diameter. With a hoop under different conditions, as regards temperature, the result would be disadvantageous, and might end in fracture, owing to the want of sufficient ductility in cold iron.

Longitudinal Strain.

In considering the strength of guns, and the causes which determine their failure, it has to be clearly understood that besides the force tending to burst them in the lateral direction—that is to say, outwards circumferentially—they are also subjected to another force which tends to destroy them in the direction of their length; the same force, which sends the shot forward, has an equal reaction on the breech of the gun, thus tending to separate it from the barrel.

The student, on considering these two distinct strains or forces, will be led to perceive that the breech of a gun is in some respects like a beam or girder; it must have a neutral surface; there must be some point where these two forces are not favourably situated for harmonious action.

If we suppose the gun to be composed of some substance resembling highly elastic india-rubber, it will be readily conceived that, under such imaginary circumstances, at the moment of explosion, the whole of the rear part of the gun would be extended in every direction, like a bladder, or as we see in the operation of glass-blowing. It is quite different, however, in the case of an iron or steel structure; there is not the same freedom for distension in every direction; and hence, when the metal is overcome, the usual result of the explosion is the blowing out of the breech, away from the barrel, and the bursting of the cylindrical part next to the breech, generally leaving the fore part of the gun entire.

The longitudinal strain, then, which comes upon the breech, is derived from the same force as that which sends the shot forward, but with this difference, however, that the gun not being so light as the shot, it recoils a proportionately less distance. The safety of a gun will depend upon the mass of solid metal composing the breech, even when the metal is merely considered to resemble an ordinary steam-hammer anvil-block, which is the instrument for receiving the force of the blow of the falling hammer; and every observing forgeman knows how efficient is the blow with a heavy anvil-block, and how well it stands up to its work, as compared with a light one. Upon the same principle, the heavier or more massive the breech of the gun, so the less is the tendency for it to recoil and break away from the cylindrical portion, by the effect due to the blow of the explosion.

For the foregoing reasons, all designers of guns who know their work introduce as much solid metal as possible, or as much as may be admissible, into the breech of a gun, merely to act the part of an anvil-block, and thus prevent the neutral surface, where the two forces unite, from being unnecessarily disturbed, or to such an extent as would endanger the safety of the gun structure.

Strengthening of Cast-Iron Guns.

The gradual dawning of the foregoing principles, on the minds of many individuals, has led to several attempts at strengthening ordinary cast-iron guns, either by exterior hooping with wrought iron, or by lining the interior, either with wrought iron or with oil-tempered steel. The former system has hitherto been unsuccessful, no doubt owing to the weak interior of cast iron giving way, before the assistance of the hoops came into exercise, and such guns have seldom done more work with the hoops, than might have been expected without their presence; hence, for the present, that system has been abandoned.

Cast-iron guns, when strengthened on the latter system —namely, by the introduction of an inner lining of some stronger material—have afforded most satisfactory results, and many guns have, therefore, been treated on this principle. Various plans have been proposed, both by Major Palliser and Mr. Parsons, but these do not differ much in regard to their general principle; the chief peculiarities of the different plans are in the details, which may here be disregarded.

The general principle upon which all such guns are strengthened is founded on the natural law, discovered by Barlow, and by him demonstrated mathematically, and which is generally assented to by all who have followed in his footsteps, and which has already been referred to in connection with the subject of built-up structures of wrought iron.

On the assumption, therefore, that when a cast-iron gun is exposed to internal pressure, the metal composing it is thereby strained unequally, if the solid gun cylinder were supposed to be divided into a number of thin concentric cylinders, the extension of each under strain would be, according to Barlow, inversely as the square of the diameter;

and as the strain is in exact proportion to the extension, then the strain will vary in the same ratio.

The interior of the gun is therefore subjected to the greatest strain, and the exterior to the least strain; consequently the interior has to stretch more than the outside. If, then, the interior is made of a stronger material and a more extensible metal than the exterior, by inserting into the cast-iron gun a tube of wrought iron or steel, the metals so arranged will be placed in accordance with the true theoretical conditions, and each portion will take its appropriate share of the work. When a discharge of the gun takes place, the inner wrought-iron or tempered steel tube will offer the first resistance and take a large portion of the strain; but, in doing so, it will stretch three times as much as it would have done had it been made of cast iron, and yet without exceeding its apparent elastic limit. But its exterior surface will necessarily stretch less than its interior, and therefore the cast iron into which it is fitted will only be stretched to the same extent, a point of importance. If the steel tube is properly fitted into the cast-iron bore, the fit being what may be termed easy, or so proportioned as to give such an amount of extension to the steel tube as will be just sufficient to stretch the cast iron up to a point a little under its elastic limit, at the moment of explosion, then the cast iron will not be injured. But these conditions imply a degree of nicety which may seem difficult to attain, by those who are not initiated, but which in reality is easily accomplished, practically, by the aid of the vernier.

A remarkable feature in connection with this part of the subject, and which the student should note, is the paradoxical fact, which can be satisfactorily proved, that by boring out a cast-iron gun to receive a steel tube, even to the extent of half its thickness—that is to say, by cutting away half its substance—the gun is thereby rendered stronger than it was before—not much stronger, but still the difference is for the better. This, however, will only hold good when the internal strain is applied on the same diameter of bore as it was originally. This is simply due to a greater harmony existing among the laminæ, because when the difference between the diameter of the enlarged bore and the external diameter is smaller than it was before, there is now less difference in the relative extensions, and consequently the various laminæ act together more uniformly, and therefore more advantageously.

Likewise the longitudinal strength of the gun is not injured, for the areas of circles being as their diameters squared, half the thickness of the gun may be bored out, and it will still retain more than half of its longitudinal strength; and as the longitudinal strain on a gun is only about one-fourth of the tangential strain, the gun will still possess an ample excess of longitudinal strength. These considerations have, therefore, led to the system of strengthening cast-iron guns by the insertion of a steel or wrought-iron lining, which has given high results, although not equal to those obtained with built-up guns of steel and wrought iron.

The student should not overlook the foregoing principles, because his pursuits may lie in some other direction. The same principles apply in numerous cases, wherever hollow cylinders of cast iron or other metal are exposed to internal pressure, of sufficient force to call their full strength into requisition.

If the study of this unpretending little volume has produced the desired effect on the mind of the earnest student, he will, doubtless, have perceived how much his daily routine duty in the workshop is associated with natural science and practical art, and how very little even of art belongs entirely to man or to man's doings; that all the materials to which reference has been made, together with all their properties, exist irrespective of man; and that the principles which determine strength in structures of any kind—beams, gearing, pillars, cranes, boilers, and even guns

—are all natural principles which have simply to be complied with or taken advantage of, in order to make the best of things as they are found in the world. There is a sort of conventional notion, especially in the workshop, that to men belongs the credit of inventing principles; those involved in the forms given to beams, for example, in order to obtain the greatest strength with a given quantity of material. But it is otherwise; men have only gradually perceived the natural principles on which strength depends, and then endeavoured approximately to comply with their requirements. Generally, natural principles can only be approximately complied with, because numerous practical difficulties present themselves in consequence of our imperfect knowledge, and bar the way to the true form in carrying out practical work. So it is in everything; the natural law, for example, upon which the mechanical device called a lever is founded, is the same principle which governs all the other so-called mechanical powers, however variously devised, and explains all our mechanical expedients for the conversion of force or motion. By no agency can mechanical work be lost or gained, however ingenious the contrivance; it can only be wasted or misapplied for the intended object, by the unnecessary friction or by the imperfect arrangements of man in accomplishing his purpose. The law itself is perfectly simple, and when the full knowledge of the law is possessed by all workmen engaged in its application, in the wide domain of applied mechanics, we are fully warranted in the anticipation that our apparatus will be gradually simplified, that the various members of complicated structures will be so proportioned to the stress which comes upon them, that a greater uniformity of strength and saving of material will thereby be achieved. So it is likewise with the most common operations of the workshop; all of them depend on some natural principle; the tempering of a chisel, the angle of a cutting tool, the shape given to a hammer, so as to give the required quality of blow, the speed suitable for the cutting of different materials, and every other kind of operation or process, all are founded upon unalterable laws of nature; our province is merely to apply materials, wherewith to direct and control the natural laws, in order to effect some definite end of our own.

From day to day, men are seeing new ways of turning these natural principles to account; it has been the same for thousands of years; these new applications are sometimes called discoveries or inventions, but all these new contrivances of the present, existed at the beginning of applied mechanics, only with this difference, that men did not perceive them. So it is now; we are groping in the dark for new appliances, but all the appliances of the future are already existing in nature, but our ignorance prevents us from seeing them at the present time; and every fresh ray of light, that shows us how to apply natural law and natural material in a new way, will never bring us nearer to the end : every improvement will only form a clue to other applications, so that invention will never cease, nor will men's work be done;—the gradual result will be to enable us to live with less expenditure of material and labour, and thus to ameliorate the condition of the whole human family.

It is scarcely one hundred years ago since it was considered next to impossible to turn or bore cast iron; now we can perceive how thin the veil was which obscured that possibility—that it was only necessary to move the surface at a certain velocity. At that period it was thought essential to take the power from the motor at a slow velocity, thus entailing heavy cumbrous mechanism to transmit the power; yet men knew the law then just as well as we do now, but they did not realise it with our vividness, so as to warrant them in changing their system; quickening the motion would at once have reduced the cost both of the power required and the materials employed.

It is not many years since it was deemed impracticable

to sharpen hard steel circular cutters, without first softening them, yet we now see that it can be done, as easily as sharpening a drill at the grindstone, and by an arrangement as old as the cutting of diamonds, and upon the same principle. The effect due to velocity has yet to play an important part in many of our modes of treating materials; the old experiment of firing a candle through a door, or of firing a pistol bullet through a pane of glass, without cracking the pane, exhibited an important principle, but the last generation could not read the lesson; by understanding this principle and applying it in other directions, refractory granite or other hard substances can be treated with ease and facility.

Again, take the steam-engine, for example. The little steam-engine which worked in the courtyard of Hero of Alexandria, two thousand years ago, was no doubt considered wonderful in its time; but if the indicator diagrams of its performance, together with the quantity of fuel consumed, per horse-power per hour, had been handed down to us, we, with our more advanced experience, would not be surprised to find the consumption of coal equal to 100 lbs.—or, going back only two hundred years, it is probable that Captain Savery's engines consumed 50 lbs.—per horse-power per hour.

Then James Watt, by fresh devices, took advantage of nature's secret working in better ways, and thus reduced the consumption to 10 lbs. of coal, and, by the further development of the same principles, the consumption of some modern engines is within a fraction of 2 lbs. of coal, per horse-power per hour. But are we to remain at 2 lbs.? We learn from Joule's equivalent that the heat which is required to raise a pound of water 1° would raise 772 lbs. 1 foot, if entirely converted into mechanical work; if so, and there is no reason to doubt, then our work with coal and steam is scarcely begun; our best engines do not perform one-tenth of that duty. Those who are now engaged in the contest are prone to think, that with 2 lbs. of coal, per horse-power per hour, we are near the limit of improvement, but let not the student think so ; we have every reason to believe that unlimited devices are stowed away in the archives of nature, which, one by one, will in due time be found by the earnest seeker, and will open up avenues of economy that are unknown in our present philosophy. The thin curtain which hides from us the means of obtaining a horse-power, with 1 lb. of coal per hour, is gradually being withdrawn, and the way of overcoming certain practical difficulties, that bar the way at the present time, is clearing up from year to year.

When the majority of men, who are engaged in the fabrication of materials into definite forms, in order to prepare them severally for some mechanical combination, for the conversion of force or motion, are all well imbued with the knowledge of natural principles, the effect must be great. There are now many thinking minds directed to the study of such questions, by the teaching of art and science, whose influence will soon make itself felt in every direction, and through all our operations.

The object of these concluding remarks is to draw the attention of the student more to the physical and experimental basis of the study of applied mechanics, force, motion, structures, and material than is usually given. The cultivation of the habit of looking for the natural laws which underlie all mechanical operations will prepare the mind for the higher and, at present, unknown practical questions that are laid up in the future.

BIBLIOBAZAAR

The essential book market!

Did you know that you can get any of our titles in our trademark **EasyRead**[TM] print format? **EasyRead**[TM] provides readers with a larger than average typeface, for a reading experience that's easier on the eyes.

Did you know that we have an ever-growing collection of books in many languages?

Order online:
www.bibliobazaar.com

Or to exclusively browse our **EasyRead**[TM] collection:
www.bibliogrande.com

At BiblioBazaar, we aim to make knowledge more accessible by making thousands of titles available to you – quickly and affordably.

Contact us:
BiblioBazaar
PO Box 21206
Charleston, SC 29413